A Beginner's Guide to Coding on iPads & iPhones

Jim Gatenby

BERNARD BABANI (publishing) LTD
The Grampians
Shepherds Bush Road
London W6 7NF
England

www.babanibooks.com

Please Note

Although every care has been taken with the production of this book to ensure that all information is correct at the time of writing and that any projects, designs, modifications and/or programs, etc., contained herewith, operate in a correct and safe manner and also that any components specified are normally available in Great Britain, the Publishers and Author do not accept responsibility in any way for the failure (including fault in design) of any project, design, modification or program to work correctly or to cause damage to any equipment that it may be connected to or used in conjunction with, or in respect of any other damage or injury that may be so caused, nor do the Publishers accept responsibility in any way for the failure to obtain specified components.

Notice is also given that if equipment that is still under warranty is modified in any way or used or connected with home-built equipment then that warranty may be void.

First Published – October 2015

British Library Cataloguing in Publication Data:

A catalogue record for this book is available from the British Library

ISBN 978-0-85934-756-3

Cover Design by Gregor Arthur

Printed and bound in Great Britain for Bernard Babani (publishing) Ltd

About this Book

Coding, or computer programming as it was formerly known, can be a daunting task for the beginner, dominated by jargon and complex new technology. Ideally, everyone in this digital age should at least have an understanding of how computers work and what they can achieve. To this end, the National Curriculum in English schools makes it compulsory for all children to study computers and coding.

My aim in this book is to explain in plain, simple English, the basic skills needed to start coding. This draws on many years experience of teaching and writing about computers. The work is based on the Python programming language, which is easy to learn yet powerful enough for demanding professional and scientific applications. Python is also one of the languages used in a new BBC project, which aims to introduce coding to millions of children via a small, programmable circuit board known as the **micro:bit**.

Many people of all ages will do coding at work or in education using laptop or desktop computers; this book shows how the best-selling iPads and iPhones, now available in many homes, can be used as a valuable tool to continue learning and practising coding.

After discussing the basic components common to all computers, the book explains how to set up an iPad or iPhone to start coding. All of the major Python programming features are then described in small, simple steps with clear examples and lots of practice exercises.

Saving and managing programs is also discussed together with copying program files between different types of computer using the "clouds" and e-mail. This allows you to continue developing a program in different locations.

About the Author

Jim Gatenby trained as a Chartered Mechanical Engineer and initially worked at Rolls-Royce Ltd., using computers in the analysis of jet engine performance. He then obtained a Master of Philosophy degree in Mathematical Education and taught maths and computing for many years to students of all ages and abilities, in school and in adult education.

The author has written over forty books in the fields of educational computing, Microsoft Windows and more recently, tablet computers. His most recent books have included "An Introduction to the Nexus 7", "Android Tablets Explained For All Ages", "An Introduction to the hudl 2" and "An Introduction to Android 5 Lollipop", all of which have been very well received.

Acknowledgements

I would like to thank my wife Jill for her support during the preparation of this book and our son David for the artwork on page 2. Also Michael Babani for making the project possible.

Trademarks

Contents

The Pythonista App

This book is based on Pythonista 1.5, from omz:software. This app, currently available in the App Store, is based on version 2.7 of the Python language.

Pythonista 1.5 can be used on iPads and iPhones running the operating systems iOS 7.0 or later.

> Any notes in this book referring only to the iPad can safely be assumed to apply also to the iPhone.

Cross Platform Compatibility

Coding or programs developed using Pythonista arc compatible with other devices such as PC computers running Python 2.7 and Android tablets and smartphones using the QPython app (but not the QPython3 app).

Screen Output

For the purpose of clarity, instead of actual screen "dumps", some listings and output have been transcribed using different background colours and text fonts.

Welcome to Python.org

This is the official Python Language Web site at:

https:www.python.org/

The Web site contains Python downloads, tutorials, documentation and library listings.

1

Introducing Computers

What is a Computer?

Mostly we may think of computers as laptops, desktops, tablets and smartphones but many other machines such as cars and household appliances also have computers built in.

All of these computers have certain features in common:

- They cannot think like human beings do.
- They follow instructions written by people.
- The instructions are stored inside the computer.
- They carry out millions of instructions per second.

Why Learn Coding?

Coding is the writing of sets of instructions for a computer. These are known as ***programs*** and are written in a special language using words from the English language. This book uses a version of the popular Python 2.7 language, delivered via an app called ***Pythonista***, designed for iPads and iPhones. Learning coding is a good idea because:

- It's an important part of the school curriculum.
- It should help you to understand computers better.
- You will learn useful problem solving skills.
- You might take up coding as a job or as a hobby.

Types of Computer

This book is mainly about coding on iPads and iPhones. If you're not too familiar with computers, the next few pages explain the main parts of all computer systems. This should help you to understand the work later in this book which involves writing your own code.

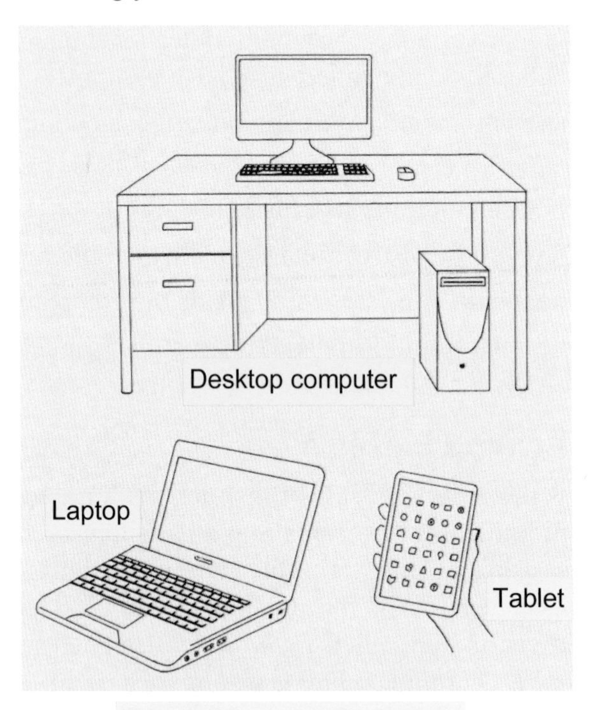

Types of Personal Computer

Coding on an iPad or iPhone

Handheld iPads and iPhones are actually powerful computers and can be used for many of the tasks done by much larger machines. ***iPads and iPhones can easily be used for coding, anywhere and at any time.***

The images on the previous page show the main types of computer in use today at home, at work and in schools and colleges. An iPhone is very similar to an iPad tablet in most respects but a little smaller.

Although the various computers on page 2 look very different in size and in their layout, they all fit the basic definition of a computer, as shown below:

Computer

A machine which can ***store instructions***, enabling it to carry out various tasks or processes, such as editing text, drawing, calculating, playing games or music, etc.

No matter what task you are doing, all computers go through the same main stages, as shown below.

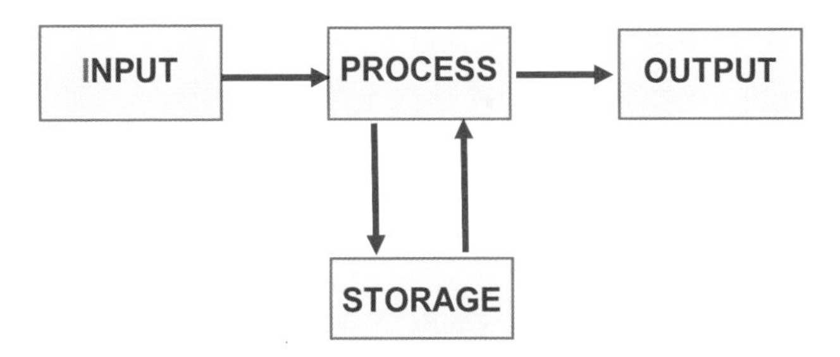

The above four stages are described in more detail on the next few pages.

The Input Stage

This is the entry of text and numbers into the computer. It may also include spoken words and data downloaded from the Internet to your computer. Common input devices are the keyboard, mouse and microphone. iPads and iPhones have their own on-screen keyboard as shown below.

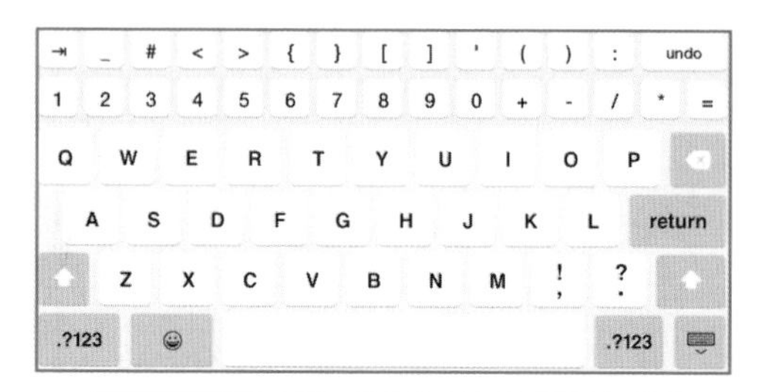

An iPad Mini on-screen keyboard

Separate physical keyboards are also available. Some people may prefer these for more lengthy coding tasks.

The Process Stage

All computers have a ***CPU*** or ***Central Processing Unit***. On small computers and tablets, etc., this is a single microchip, as shown below. The CPU or ***microprocessor*** is often called the "brains" of a computer because it carries out all the instructions, calculations, etc. The CPU carries out millions of instructions per second, measured in GigaHertz or GHz for short. The latest iPads and iPhones have CPU speeds of around 1.3-1.5 GHz, so they can work just as fast as many bigger computers, such as laptops and desktops.

A CPU chip or microprocessor

The Output Stage

This presents the results of the program currently being ***run*** or ***executed***. The output is commonly displayed on the screen in the form of text or a game or a photograph, as shown below. Output may also be printed on paper, such as an essay, magazine or newsletter. Other forms of output include music, video and TV and radio programmes.

A photo as output on an iPad

Storage of Programs and Data

Programs

As mentioned earlier, programs or code are the instructions in the Python language, telling the computer what to do.

Data

Data, often called ***raw data***, are the recently collected facts and figures you ***input*** into a program such as names and addresses, ages, weights, heights, temperatures, etc.

Information

After processing raw data in the computer, you should produce ***meaningful information*** as ***output***, such as average temperatures in summer or graphs to show rainfall.

Internal Storage (Not the Same as Memory)

Programs and data are recorded ***permanently*** on the ***Internal Storage***, also known as ***backing storage***. This has a similar role to the hard drives on laptop and desktop computers. Once you've saved programs and data on the backing storage you can retrieve and use them whenever you need to. Otherwise you would need to keep repeatedly typing in all the same words and numbers.

The Internal Storage inside an iPad or iPhone, is a form of ***flash memory*** similar to the technology used in a removable ***flash drive***. Unlike the hard drive, which rotates at high speed, the flash memory has no moving parts. The flash memory is soldered to the circuit board on an iPad or iPhone and cannot easily be upgraded. This differs from the ***SSDs (Solid State Drives)*** used in many tablets and smartphones.

The SSD also uses flash memory but is enclosed in a case and connected by cables, so it can be upgraded if needed.

When you switch the computer off, programs and data will remain on the backing or Internal Storage. However, you can also ***delete*** from the backing storage any programs and data you no longer need.

Depending on the model, iPads and iPhones can have Internal Storage ranging from 16GB to 128GB, compared with 500GB or 1000GB in a laptop or desktop computer. (Terms such as GB are discussed on the next page.)

The fact that an iPad or iPhone has much less Internal Storage than a laptop or desktop machine is not a serious problem. Tablets and smartphones can store most of their files such as photos and documents in the ***clouds*** on the Internet. (The clouds are really big Internet computers provided by Apple, Google and Dropbox, etc.)

You can also connect external storage media such as SD cards from cameras, to import photos to the iPad or iPhone and save them on the Internal Storage.

The Memory or RAM

This is ***temporary*** storage which is cleared or wiped when the computer is switched off. Programs and data which you currently wish to use have to be fetched from the backing store and placed in the ***memory*** or ***RAM*** (***Random Access Memory***), from where they are ***run*** or ***executed***.

The RAM is sometimes called ***volatile*** storage, while the permanent backing storage is said to be ***non-volatile***.

iPads and iPhones typically have 1GB of memory or RAM, while the latest iPad Air has a more generous 2GB.

Units of Storage

Both backing store and the memory can hold billions of letters and numbers. The main units of storage are:

Byte:	For example, the space needed for one letter.
Kilobyte:	1024 bytes.
Megabyte:	1024 kilobytes or about a million bytes.
Gigabyte:	1024 megabytes or about a billion bytes.
Terabyte:	1024 gigabytes or about a trillion bytes.

Hardware

This means all the physical parts of a computer system, including the screen or monitor, the casing, the processor and all the circuits and microchips and other electronic components. The hardware also includes any printers and other ***peripheral*** devices such as separate keyboards and mice, which can be used with tablets as well as laptop and desktop computers.

Small hand-held computers like iPads and iPhones don't have the bulky hardware found on bigger computers. However, very small versions of these components, such as speakers and microphones, are still present, in an iPad or iPhone.

Various adapters are available which enable external devices such as USB flash drives and SD camera cards to be connected to an iPad or iPhone, e.g. for the importing of photographs.

2

Computer Software

What is Software?

Unlike the ***hardware*** just discussed, software is not made up of physical parts that you can touch. Software means all the ***programs*** or ***sets of instructions*** consisting of words and numbers, saved on the Internal Storage, i.e. ***flash memory***, of an iPad or iPhone. There are two types of software, ***systems software*** and ***applications***.

Systems Software

The Operating System

This controls the overall running of a computer, managing tasks such as the screen display, the saving of programs and data and providing the ***Graphical User Interface*** (***GUI***). The GUI is the system of icons or small pictures and buttons on the screen used to launch apps, etc. The operating system used on iPads and iPhones at the time of writing is ***iOS 8.4***, with ***iOS 9*** due to be launched shortly. ***Android***, produced by Google, Inc., is another major operating system used on many tablets and smartphones.

Larger Apple computers such as the MacBook and iMac currently use ***OS X Yosemite*** while ***Microsoft Windows*** is the operating system used on most laptop and desktop PCs.

The operating system is normally already installed on the Internal Storage of a brand new computer. New versions of an OS can usually be freely ***downloaded*** to a tablet or smartphone from the Internet.

Device Drivers

Device drivers are small programs used to enable accessories, such as printers, etc., to work with your particular operating system, such as iOS or Windows.

Utilities

These are programs used to help with the running, maintenance and security of the computer, such as a ***virus checker*** or a ***debugging*** utility for correcting errors in programs. Some utilities are built into the operating system.

Applications Software (Apps)

The systems software just described is needed whatever you are using a computer for. The programs you want to run for your own work or entertainment are known as *Applications software*. These might include a game, a drawing program, photo editor or a word processor, for example. Some apps are usually already installed on a new computer but you can obtain more apps and ***install*** them, i.e. save them on the *Internal Storage*. On laptop and desktop computers, new applications software is often supplied on a CD/DVD or downloaded from the Internet. Then it must be permanently saved on the Internal Storage such as a ***hard disc drive***. On tablets and smartphones new apps are usually downloaded from Internet storage such as the *App Store*.

The App Store

There are millions of apps to choose from, to download and install to your iPad or iPhone. These include games, videos, music, business and photo editors as shown in the small sample at top of the next page.

Best New Apps

Tiny Builders - Digger, Crane...	Rookie Cam - Photo Editor...	Laundrapp – Your Dry Clea...	Bee - Email Smart and Fast
Entertainment	Photo & Video	Utilities	Business
£1.99			

Pythonista

If you type **Python** into the Search bar in the **App Store**, a number of apps are displayed, including notes on the Python language and also an eBook on the Monty Python comedy series and films, after which the Python programming language is named.

One of the apps displayed, ***Pythonista***, is a complete programming system for the Python 2.7 language. Pythonista is used throughout the rest of this book to demonstrate the basic skills of programming in the Python 2.7 language.

Once installed on your iPad or iPhone you will be able to:

- Use the ***Console*** to test short Python 2.7 commands ***interactively***.
- Use the ***Editor*** to ***code***, ***save***, ***run*** or ***execute*** and ***edit*** programs in the Python 2.7 language..
- View and run sample programs from the ***scripts library***.

Planning a Program

Many tasks can be broken down into a number of tasks to be carried out in a certain order. For example, take a simple task like watching a television programme. The steps might be as follows:

Switch on the TV
Select the programme
Watch the programme
Switch off

Normally you might want to watch another programme before switching off, so we can represent this better with a flowchart, as shown below:

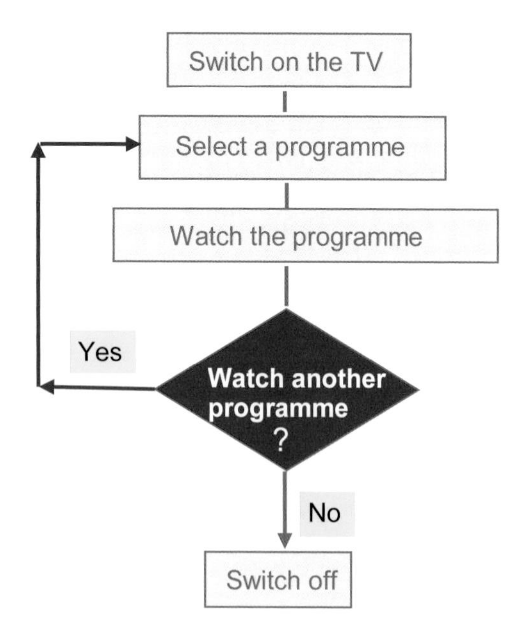

Algorithms

The flowchart on the previous page is a simple example of an ***algorithm***. This is a sequence of steps to solve a problem. A computer program may be made up of many algorithms to tackle different problems.

Decisions and Branching

The flowchart on page 12 introduces another important feature of many programs. This is a ***decision***, shown here in the red diamond with a question mark. It is also known as a ***branch***, because we can proceed in one of two directions, depending on the answer to the question.

Loops

If the answer is Yes, then we ***loop*** back and select another TV programme. If the answer is No, we continue down and switch off. This example introduces three important ***reserved words*** used in Python coding, if, else and while.

if and else

The decision on the flowchart on page 12 is really saying if something is True then do one thing, else if not true, i.e. false, do something different.

while

The procedure branches back if you want to watch another program and you could do this many times if you wish. This is a ***loop*** which allows the ***repetition*** of a task over and over again. You would do this while you still wanted to watch television. while is another important reserved word in the Python language and allows you to keep repeating some steps as long as something is True.

A Maths Algorithm

The example below shows you the steps to display the numbers up to 50 in the form of a simple algorithm.

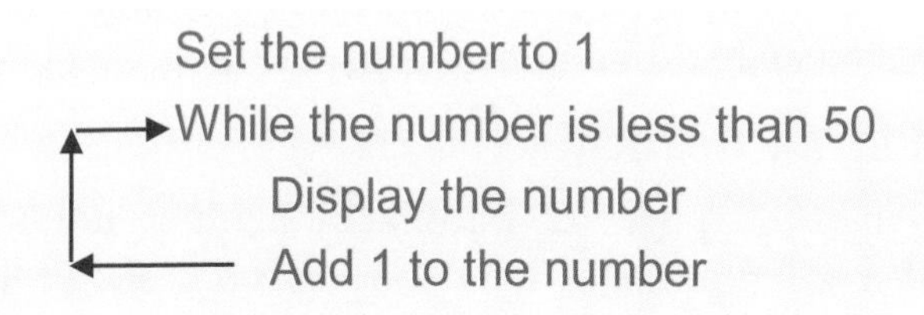

This would be coded in the Python language as follows:

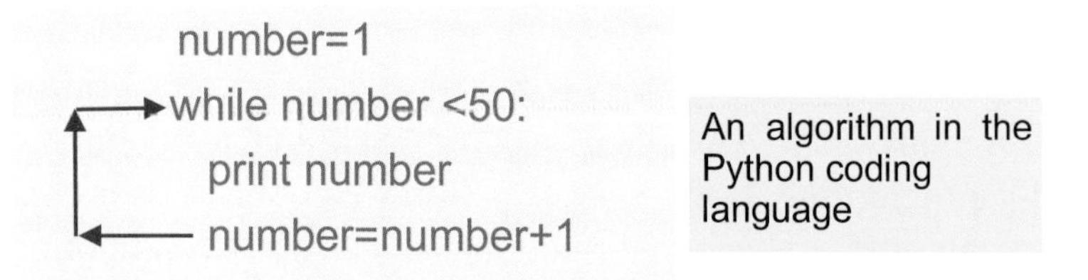

An algorithm in the Python coding language

(print in Python causes the output to appear on the screen).

Chapter 3 starts coding in detail, so don't worry if you don't yet understand the Python code above. If you do understand, perhaps you could rewrite the algorithm to print only the odd numbers from 1 up to 100.

Exercise

Write some simple algorithms in ordinary English for some common tasks. Try to include a decision and a loop. Here are a few examples to get you started:

- Mending a puncture on your bike.
- Preparing a meal.
- Preparing for a trip or holiday.

Inside the Computer

The digital computer, being an electronic device is a ***two-state*** system. This can be thought of as, say, an electronic pulse flowing or not flowing or a row of light bulbs switched ON or OFF, as shown below.

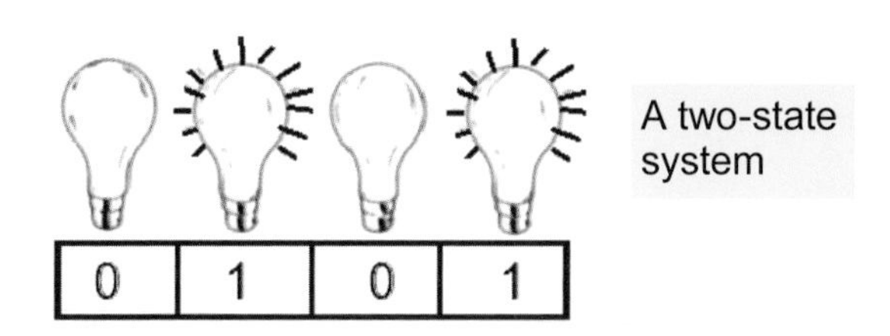

A two-state system

The two-state system can be used represent the two digits 0 and 1, as shown above. So everything at the very inside of the computer has to be represented by a pattern of 0's and 1's, known as the ***binary code***. These 0's and 1's are known as ***binary digits*** or ***bits*** for short. They are normally arranged in groups of ***8 bits*** known as a ***byte*** (whereas a group of ***4 bits*** is called a ***nibble***).

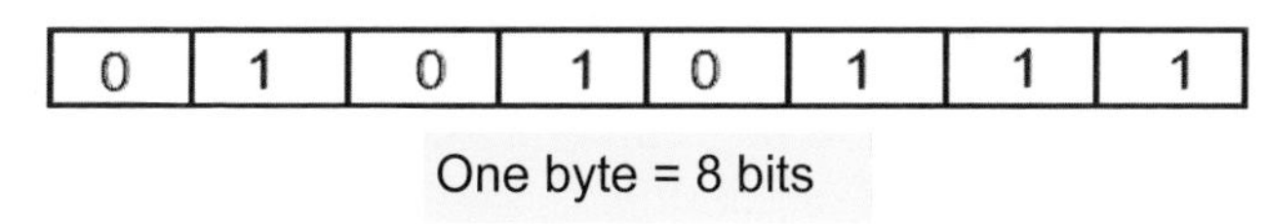

One byte = 8 bits

The byte can be thought of as a set of storage boxes which can represent :

- A ***keyboard character*** such as a letter, digit 0-9, etc.
- An ***instruction***, e.g. to add two numbers.
- A ***number*** such as 19,567.
- An ***address*** of a storage location in the memory.

High Level Languages

The computer, being an incredibly high speed device, has no problem in manipulating the long strings of binary digits. Humans, however, don't have the time or patience to feed the computer with long strings of 0's and 1's. So computer scientists invented ***high level languages***. These are much closer to English, using words such as print, if, else, while, input and many more.

Python 2.7 is one of the most popular high level languages because it's powerful yet easy to learn. ***Pythonista***, used throughout this book, is an app which allows you to run the Python 2.7 high level language on iPads and iPhones.

Python Scripts

Python is known as a ***scripting language*** and the programs you write, known as ***scripts***, are saved as ***files*** with the ***.py*** extension, such as ***mygame.py***. Pythonista scripts can be written and saved using the built in ***Editor***.

The Interpreter

To ***run*** or ***execute*** a Python script, it has to be translated line by line, into the machine's own binary or machine code. This translation process must be done every time you run the program. Translating a Python script, which uses words like print and while, for example, is done by a program called an ***interpreter***. Pythonista has a built-in interpreter.

The Compiler

Unlike the scripting languages such as Python, some high level languages, for example Fortran, take the code or program written by the user and translate it all into a standalone file in the machine's own binary code. This file can be run whenever needed without any further translation.

Coding and Running Programs

Instead of installing and using programs that other people have written, the following pages show how you can download Pythonista and start coding your own apps. To begin with you can type in the commands, such as print, at the keyboard. These can be run in ***interactive*** or ***immediate mode*** and produce output straightaway, for example, to print your name on the screen. However, for longer programs you need to:

- Type in the code, i.e. instructions.
- ***Save*** the instructions permanently on the Internal Storage, i.e. flash memory on the iPad or iPhone.
- ***Fetch*** i.e. retrieve the instructions from the backing store to the memory or RAM whenever needed.
- ***Run*** or ***execute*** the program from the RAM.
- When finished, close the program and shut down.

When the computer is switched off, the program, i.e. instructions, will be wiped from the memory or RAM but will still be permanently saved on the Internal Storage, i.e. backing store. So the program can be reloaded from the backing store and run whenever you want to in the future.

This book uses the Pythonista app to introduce basic skills for the Python 2.7 language on an iPad or iPhone. You might also wish to use Python 2.7 via a different app or on a different computer system altogether. If so, the rules and ***syntax*** of the Python 2.7 language which you learn from this book will still apply.

Installing Pythonista from the App Store

Tap the App Store icon shown on the right and then search for **Pythonista**. You will then see the Pythonista icon, price, etc., as shown below.

Tap the price (currently £4.99) and then tap **BUY**. Next enter your Apple ID and password and tap **INSTALL**. After a short time, Pythonista will be installed on your device and the icon shown below on the right will appear on your Apps screen.

You are now ready to start coding using the Python language, as discussed in Chapter 3 onwards. You can launch, i.e. open, Pythonista at any time by tapping its icon, shown on the right, on your Apps screen.

Please Note:

Pythonista 1.5 (the current version in the App Store) requires iOS 7.0 or later. You can check your iOS version by tapping **Settings**, **General**, **About** and looking at **Version**. If you have an earlier version of iOS you may be able to update it with a free download “over the air”. (Tap **Settings**, **General** and **Software Update**).

3

Starting to Use Python

Introduction

Chapters 1 and 2 described the main hardware and software features of all computers. Chapter 2 also showed how to install the Pythonista app on an iPad or iPhone. In this chapter you will learn how to launch the Pythonista app and start writing simple Python code.

As described earlier, Python uses English words like print, while, if and else and these are the same when using Python 2.7 on other types of computer. This means the skills you learn for coding on an iPad or iPhone will also be useful when using, say, a laptop or desktop PC or an Android device. So you will be able to transfer your code and continue developing and running programs at home, at work or at school or college. All high level languages like Python use a fixed set of ***keywords*** or ***reserved words*** such as print, for, while, if, else, etc.

In this book, "Python" refers to version 2.7 of the Python coding language. Pythonista is the name of an ***app*** which implements Python 2.7 on an iPad or iPhone. Pythonista has a built-in ***interpreter***, which translates instructions in the Python code into the machine's own binary code of 0's and 1's. Pythonista also includes a ***script editor*** for ***coding*** (i.e. writing), ***saving***, ***executing***, (i.e. ***running***) and ***editing*** Python ***programs***, also known as scripts.

The Pythonista Screens

With the Pythonista app installed as described on page 18, tap the icon shown on the right on the Apps screen. This opens Pythonista, with four alternative screens — the ***Script Library***, the ***Documentation***, the ***Console*** and the ***Editor***. Each of these screens can be viewed after swiping horizontally inwards from the left or right of the screen.

The Script Library

The Documentation
(Explains the various screens)

The Console

The Editor

The Script Library

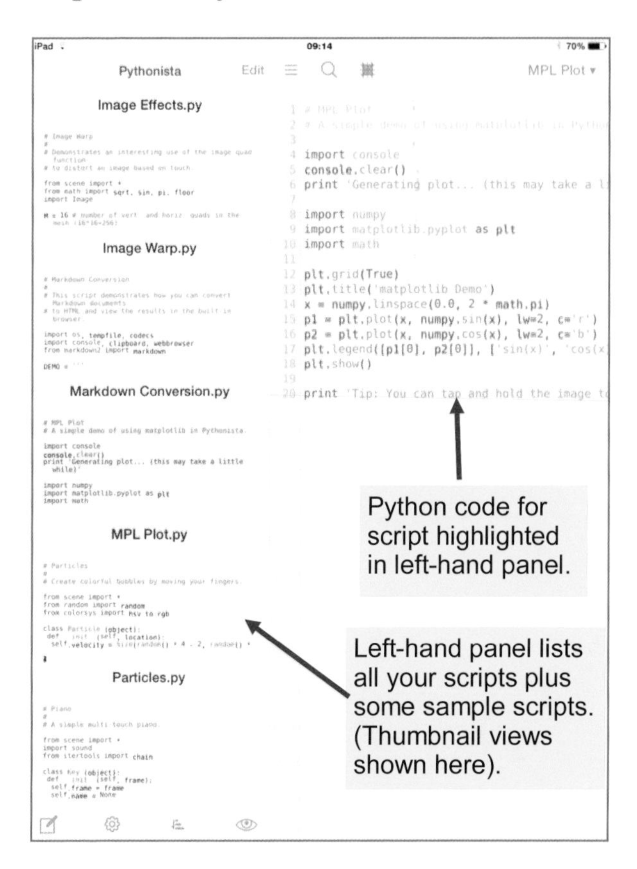

The left-hand panel above lists the scripts you've written and also some sample scripts provided in Pythonista.

The **Edit** button at the top above is used to **Delete** and **Move** highlighted scripts and also to create a **New Folder**, as discussed in more detail in Chapter 5.

Delete... Move... New Folder...

The **Script Library** screen has a small menu bar at the bottom left, as shown below and on page 21.

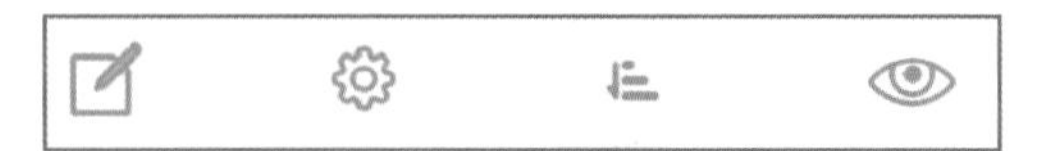

The icon on the left above and on the right is used to start a new script in the **Editor**. There are several advanced options for new scripts, but beginners new to coding should choose **Empty Script**.

The gear icon allows you to change various settings on the **Editor** screen, such as 6 alternative **Color Themes**, the text **Font** and **Font Size** and **Indentation** (discussed in more detail in Chapter 5).

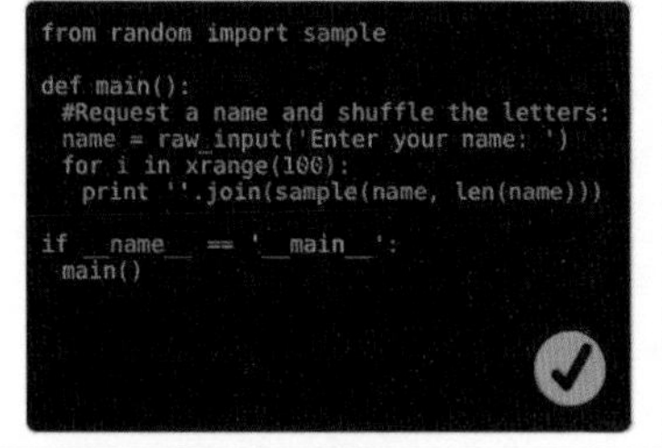

Alternative Color Themes for the Editor

The icon on the right and on the menu bar above allows you to *sort* the list of scripts in the left-hand panel shown on page 21, into date or alphabetical order.

The icon shown here on the right and on the menu bar above, allows you to switch between **Thumbnail Previews** of the scripts as shown on page 21 and a **List** of just the names of the scripts.

The Console

Swipe left twice from the **Script Library** to display the **Console**. As discussed on the following pages, the **Console** is used for entering single commands such as print "hello".

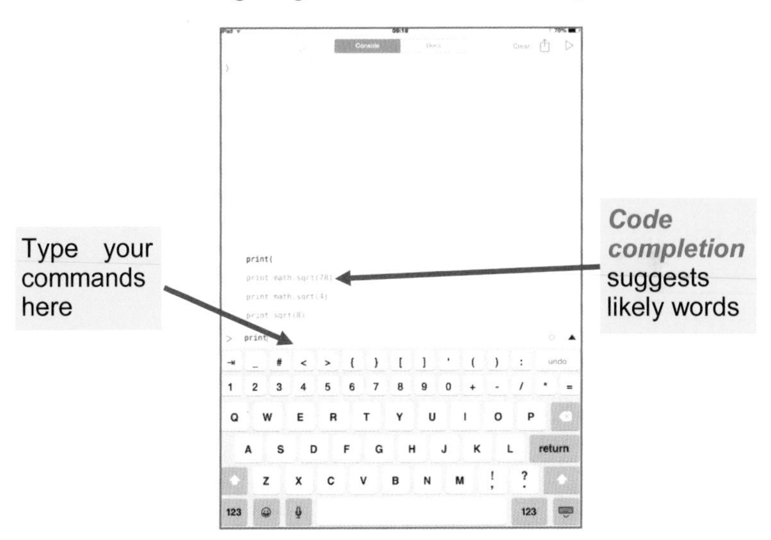

Tap in the bar at the bottom of the screen to bring up the keyboard. Then start typing a command, such as print "hello", in the bar above the keyboard. To ***run*** or ***execute*** the command tap **return** on the on-screen keyboard. The output from this simple command is shown below.

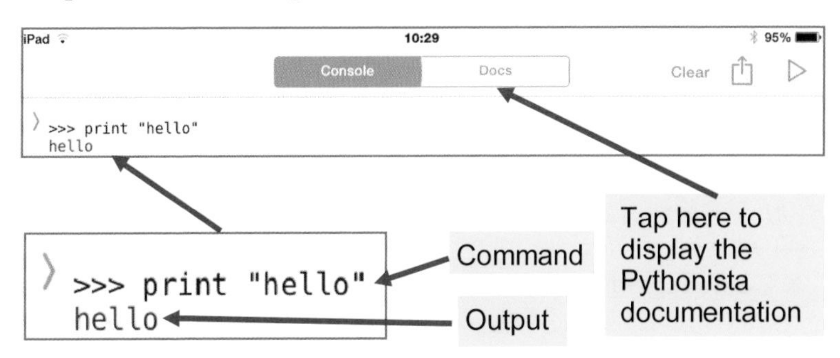

Using the print Command in the Console

This is a Python command which displays output on the screen. Text must be enclosed in quotes, as in:

Output ↘

```
>>> print "hello"
hello
```

← Command

Lower Case Letters

Python commands such as print always use ***lower case***, not capital letters. So PRINT or Print will cause the command to fail and a *syntax* error message will appear, such as:

```
'Print' is not defined
```

Spelling

Words like print and other Python ***keywords***, must always be spelt correctly. Otherwise the command will fail and an error message will appear such as:

```
'plint' is not defined
```

Speech Marks or Quotes

Words to be displayed on the screen must be enclosed in speech or quotation marks. Either single or double quotes can be used, so both "hello" and 'hello' are correct in Python, as shown below:

```
>>> print "hello"
hello
```

```
>>> print 'hello'
hello
```

However, you can't mix double and single quotes around words to be displayed. As shown below, this results in a ***Syntax Error***, i.e. a mistake in the Python grammar.

```
>>> print "hello'
Syntax Error
```

Triple Quotes

By enclosing a string of text in triple quotation marks, you can display several lines on the screen in any layout you choose.

It doesn't matter whether you have 3 double quotes or 3 single quotes as long as they are the same at both ends of the piece of text.

```
>>> print"""
He followed her to school one day
Which was against the rule
It made the children laugh and play
To see a lamb at school
"""
```

This produces the following output on the screen. (The print command used throughout this book, in this context really means "display on the screen").

```
He followed her to school one day
Which was against the rule
It made the children laugh and play
To see a lamb at school
```

Exercise: Copy and run the print command above, using triple single or triple double quotes, the same at both ends. Then repeat with a few lines of your own. Don't put spaces within each set of triple quotes.

Spacing

Adding an extra space *before* print will give an error with the message unexpected indent. As discussed in Chapter 5, ***indentation***, i.e. spaces at the beginning of a line, have a special purpose in Python. You can, however, add spacing *within* lines of code to make it more readable, as shown below.

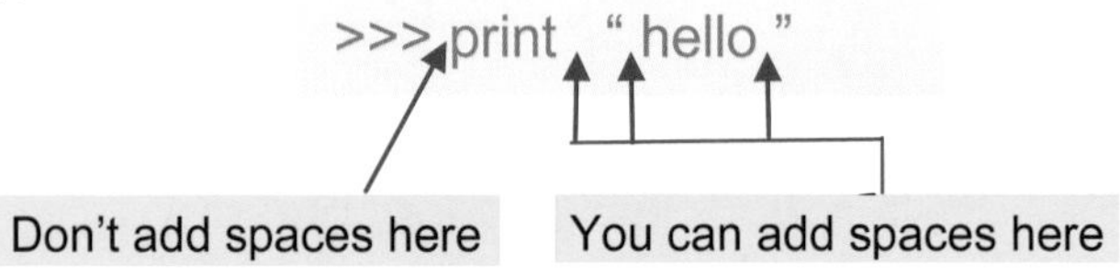

Repeating a print Command Using *

Enter the following at the command prompt in the **Console**. Again it will help if you put some spaces within the line. This separates the 4 hellos shown below.

```
>>> print "hello  " * 4
```

After you tap **return**, the output appears as shown below:

```
>>> print "hello   " * 4
hello  hello  hello  hello
```

As shown above, * 4 means repeat the print 4 times.

Exercise:

Use the above method to print your name 6 times. Put some spaces before the closing speech marks to separate each display of your name. Make sure you type print in lower case letters.

Splitting a String of Text Using \n

\n is known as an ***escape sequence*** and can be used with print to display part of a string of text on the next line, as in:

```
>>> print " The rain in Spain \n stays mainly in the plain"
 The rain in Spain
 stays mainly in the plain
```

Now enter \n before hello ***inside the speech marks***, as shown below. This displays each word on a new line:

```
>>> print "\n hello " * 4
hello
hello
hello
hello
```

The backslash "\" appears on the ***symbols keyboard*** which is displayed after tapping the key shown on the right. This key is found near the bottom left and right of the numeric keyboard.

#+=

Exercise

- Select the **Console** and enter a command to display some text such as 'Welcome to Python'.
- Tap **return** to output the message on the screen.
- Use \n to split a long sentence into two lines.
- Use print and \n to display your name 8 times, using a new line for each display of your name.

Variables

Data is held in the computer's memory in ***store locations***, just like small boxes with labels on the outside, such as.

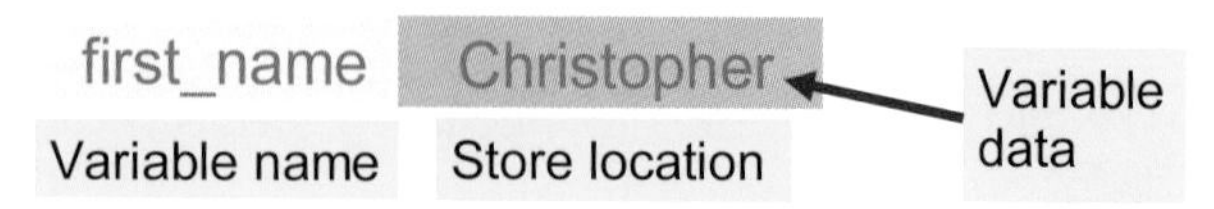

The data in a store location can be overwritten, e.g. by entering a new first_name, so the store is called a ***variable***.

Variable Names

We make up our own names for variables, such as first_name above.

- Variable names are usually mostly letters.
- You can include digits 0-9 ***within*** a variable name.
- A variable name can't start with a digit.
- You can't include spaces or Python keywords.
- You can include ***underscores*** to improve readability, as in first_ name.
- A variable name can include upper and lower case letters as in myAddress, to improve readability.

Meaningful Variable Names

You can use a single letter as a variable name such as a shown on the left below, but more meaningful names such as age make it easier for other people to understand.

```
>>> a = 17
>>> print a
17
```

```
>>> age = 17
>>> print age
17
```

String Variables or Strings

String variables or ***strings*** contain letters and keyboard characters and must be enclosed in quotes, such as:

```
surname = "Jones"
```

This line ***assigns*** the data "Jones" to a store called surname. Type the following at the command prompt, but insert your own name in the quotation marks:

```
>>> surname= "Jones"
>>> print surname
Jones
```

Now, without clearing the above commands, enter the following at the command prompt:

```
>>> surname ="Walker"
>>> print surname
Walker
```

The store called surname now contains Walker. You normally assign an initial value or contents to a variable, such as Jones above. This remains in the store until it's ***overwritten*** by the input of fresh contents.

You can also assign ***multiple variables*** in a single command, as in:

```
>>> name1, name2, name3 = "Tim", "Sue", "Pat"
>>> print name1, name2, name3
Tim Sue Pat
```

Note the commas and quotes above. You can insert spaces ***within*** the commands but not at the very beginning.

Numeric Variables

The = sign is used in computing to ***assign*** an initial value to a store, as in:

```
number=1
```

In computing, the = sign does not mean "equal to" or "the same as", as it does in normal arithmetic, such as 6=4+2.

Computers often use lines like:

```
number=number+1        or        number+=1
```

The above lines both mean: "Let the store we have called number now contain the initial value of number plus 1."

You can easily check this by typing a few commands in the **Console**, as shown below.

```
>>> number = 1
>>> number = number + 3
>>> print number
4
>>> number = number + 5
>>> print number
9
```

```
>>>number = 1
>>>number +=3
>>>print number
4
>>>number+=5
>>>print number
9
```

Exercise:

Type the commands shown above into the **Console**, tapping **return** at the end of every line. You should see that the variable number which originally contained 1, now contains 9.

Now make up 3 different examples of your own. Make up a different name for the variable store, instead of number and add or subtract various numbers.

4

Calculations and Decisions

Introduction

The previous chapters showed how you can enter ***one-line commands*** straight into the **Console** and get the results on the screen immediately. Chapter 5 shows how you can use the Pythonista **Editor** to create a ***program*** by entering and ***saving*** a list of commands, available for future use.

This chapter shows how the Python language can be used to do arithmetic. Although the basic calculations are done in a similar way to our everyday arithmetic, computers generally use some different signs for certain ***operations*** such as multiplication and division.

As discussed on the next page, computers also make use of some ***operands*** which we don't use in everyday arithmetic. These include the ***modulus*** or ***remainder*** and the ***integer*** or ***whole number***, where the part to the right of the decimal point is ignored. So for example, 7.534 in integer form would just be 7 without a decimal point.

This chapter also discusses some important ***conditions*** such as ***greater than*** and ***less than*** and True and False. These can be used to make decisions, such as:

- If your age is greater than 17 years (True) you can learn to drive a car on the roads.
- If it's a nice day (True) you might go for a bike ride.
- Else if it's not (False) you might do some coding.

Using the Console as a Calculator

If you enter a simple sum such as 9+11 at the command prompt as shown below, the answer immediately appears when you press **return**.

```
>>> 9+11
20
```

Computers use * for multiply and / for divide

So we could enter, say, 9+7*8 and get the answer 65.

```
>>> 9+7*8
65
```

Or enter (9+7)*8 and get the answer 128.

```
>>> (9+7)*8
128
```

The different answers 65 and 128 obtained above are both correct. This is because the brackets above in (9+7)*8 change the order in which the steps are carried out. The computer carries out the steps in the same order as used in normal arithmetic. One way of remembering the sequence of steps is ***BOMDAS***. This is explained on the next page.

Exercise: Type in the two examples below and note the answers. Make up 3 similar examples of your own.

```
>>> 12-7+ 8*12/4          >>> (12-7+ 8)*12/4
```

BOMDAS

B:	Brackets, also known as ***parentheses***
O:	Orders (squares, cubes, square roots, etc.)
MD:	Multiplication and Division
AS:	Addition and Subtraction

The above list means any brackets are worked out first, followed by any orders, then any multiplication and division, then finally any addition and subtraction. Multiplication and division are equal in status, so if both occur on a line, work from left to right. Similarly for addition and subtraction.

Some of the common arithmetic signs or ***operators*** used on computers are :

+	addition	7+5 ==12
-	subtraction	9-6 == 3
*	multiplication	5*6 == 30
/	divide	18.0/4.0 == 4.5
//	divide (integer)	18//4 == 4
%	remainder	21%5 == 1
**	exponent	2**3 == 8

Computing mathematical operators

As shown above, computers use some different signs compared with those used in everyday arithmetic. These differences are explained on the next page.

Computers use the following signs:

* means multiply and / means divide.

== means ***equals*** or ***the same as***, instead of = .

= is used to ***assign*** numbers, words and characters to a ***variable***, i.e. or memory store, as discussed on page 28.

In addition to the common maths operations of addition, subtraction, multiplication and division, the table on the previous page also includes the following:

//	divide (integer)	18//4 == 4
%	remainder	21%5 == 1
**	exponent	2**3 == 8

Integers and Floating Point Numbers

An ***integer*** is the ***whole number*** part of the answer to a division sum, such as 3 in the example below. You can check these by typing a few examples into the **Console** in interactive mode, as discussed below and earlier.

```
>>> 19.0/6.0
3.166666......
```

Normal division using /

```
>>> 19//6
3
```

Integer division using //

A number with figures to the right of the decimal point such as 3.166666... above is known as a ***floating point*** number or simply as a ***float***.

Remainder or Modulus

The remainder is the whole number left over after a division involving two whole numbers, e.g. 14 divided by 5 goes twice remainder 4.

```
>>> 14%5
4
```

Type % after tapping this key on the numeric keyboard **#+=**

Exponent

The exponent is the same as the ***orders*** on page 33.

In everyday arithmetic 2^3 means 2x2x2 or 8.

In this example, 3 is the exponent and tells you that 2 has to be written down 3 times and multiplied by itself.

So 2^5 means 2x2x2x2x2, for example.

In Python this would be written as 2**5.

Typing this into the **Console** in interactive mode produces the following:

```
>>> 2**5
32
```

Including Text With Calculations

Enter the following into the **Console**:

```
>>> print “9 times 5  = ”, 9*5, “   9 plus 5 = ”,9+5
9 times 5 = 45   9 plus 5 = 14
```

Insert space

Please note in the above example, the use of speech marks and commas. You can insert spaces, e.g. around = within the speech marks, to improve readability.

Making Decisions

These involve the greater than (>) and less than (<) signs shown on the next page.

When you enter, for example, 5>3 and 6<2 into the **Console** and tap **Enter**, the following results appear:

```
>>> 5>3
True
```

Is 5 is greater than 3?
Yes

```
>>> 6<2
False
```

Is 6 less than 2?
No

You can see that when asking the simple questions above, the computer answers either True or False, whereas we might answer Yes or No.

True and False are used a lot in programs. Although the computer can only use 1 and 0, this is enough for it to make a decision because 1 can be used to represent True and 0 can represent False.

For example, we might decide that if the weather is fine, to go for a bike ride, else if not we might do some coding.

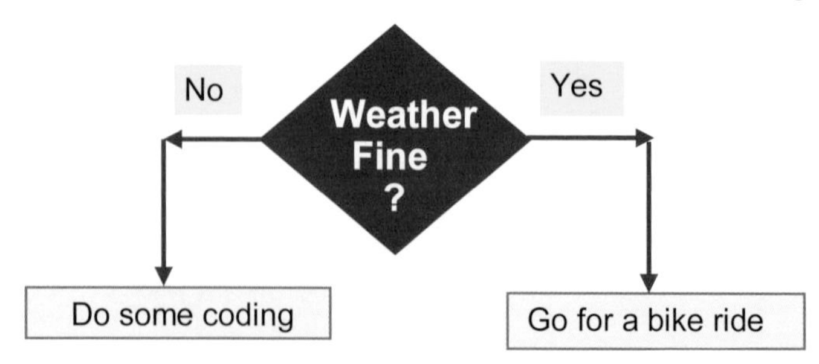

As discussed later, these Yes/No or True/False decisions can be coded using the Python reserved words if, else and elif.

Equalities and Inequalities

The following signs are used by Python:

>	greater than
>=	greater than or equal to
<	less than
<=	less than or equal to
!=	not equal to
==	equal to

Please note that = and ==have special meanings in Python, as discussed on page 34.

As shown at the top of the previous page, you can test these conditional operators using interactive mode in the **Console**.

For example:

```
>>> 7 >= 6
True
```

```
>>> 8 >= 6+2
True
```

```
>>> 5 <= 3
False
```

```
>>> 5 <= 3+4
True
```

```
>>> 6 != 3+4
True
```

```
>>> 6 != 4+2
False
```

```
>>> 7*4==28
True
```

```
>>> 18//7==4
False
```

Mixing Text and Arithmetic

You can mix text and calculations in one command, as shown below. Type the following command into the **Console**.

```
>>> print "9 divided by 2 = ", 9//2, "remainder ", 9%2
```

The output is:

```
9 divided by 2 = 4 remainder 1
```

In this example we use =, not ==, within the speech marks above as it's not being used in a computer calculation.

- 9//2 gives the integer ***quotient*** when 9 is divided by 2.
- 9%2 gives the remainder when 9 is divided by 2.
- You can include extra spaces within the speech marks to make the output easier to read.
- Note the commas in the top command between the text in speech marks and the calculations.

Exercise:

- Make up 12 examples of your own like those on the bottom of page 37. For each of the six signs (greater than, etc., at the top of page 37) make up one True example and one False.
- Write and test, in the **Console**, a print command, as shown at the top of this page, to divide 17 by 3 and output on the screen the integer quotient (i.e. answer) and the remainder, also known as the ***modulus***.

Using the Python raw_input( ) Function

This function is used to ask a user to enter some information. You can practise using raw_input( ) in interactive mode, i.e. using the **Console**. To start with, a simple text example is given, as shown below. Enter the commands shown in the blue boxes. The text prompts in the cream boxes below appear on the screen automatically.

```
firstName = raw_input ("Please enter your first name: ")
```

The above line causes the text in quotes to be displayed, as shown below. Insert a space between first name: and ").

```
Please enter your first name:
```

The computer waits for the user to enter their first name and tap **return**. This assigns whatever is typed, such as Christine in this example, to the variable store firstName.

The next line, shown in blue below, prints the text in the quotes, followed by the contents of the store firstName.

```
print "Pleased to meet you  ", firstName
```

The output that appears on the screen is shown below.

```
Pleased to meet you Christine
```

firstName or first_ name can be used for readability. If you miss any of the brackets or quotes shown above, the commands will fail and the message syntax error will be displayed. You must place a comma in the line print "Pleased to meet you ", firstName. Insert spaces within the quotes to improve readability.

Using raw_input( ) with Numbers

There is a snag when using raw_input( )with numbers.

Enter the following into the **Console**:

```
>>>num1=raw_input ("Enter first number ")

>>>num2=raw_input ("Enter second number ")

print "Total = ", num1 + num2
```

After entering the first line above, tap **return**. Then type in a number in response to the prompt "Enter first number". Then repeat for the second number. Then enter the print statement. So if we entered, say, 17 and 21 we should see:

```
Total = 38
```

Instead we see the **wrong answer**:

```
Total = 1721
```

The reason this is wrong is because the raw_input( ) function on its own treats numbers as ***strings*** of characters, not mathematical numbers. For example, a telephone number such as 07954321 is just a string of characters, not a mathematical number such as 347, which means 3 hundreds, 4 tens and 7 units.

So treating 17 and 21 as strings we get the following:

```
Total = 17+21 = 1721
```

Here the computer has wrongly used ***concatenation***, which is used to join together strings consisting of letters.

Using int( ) to Convert Strings to Numbers

This can be used, as shown below, to convert the strings produced by the raw-input( ) function to numbers.

```
>>>num1=int(raw_input (“Enter first number “))

>>>num2=int(raw_input (“Enter second number “))

print “Total = “, num1 + num2
```

Using the int( ) function as shown above prevents the string concatenation problem discussed on page 40. This allows the user to use raw-input( ) for calculations with numbers.

Spacing

Adding spacing can make the commands and output on the screen easier to read. For example, in the above code you could add one or more spaces between the word number and the quotes. You might also add some spaces around = in “Total = “. Similarly you can add spaces around the + sign in num1 + num2.

Exercise:

- Enter the above code into the **Console** in interactive mode. Make sure all the brackets and the speech marks are copied exactly, plus the comma in print “Total = “, num1 + num2. Make sure the total is correct.
- Using the **Console** in interactive mode, make up a similar set of commands to add 4 numbers. Make up your own variable names instead of num1, etc. and your own prompts within the quotes.

Key Points: Console/Interactive Mode

- Allows you to enter one-line commands.
- The *interpreter* translates the commands and returns the output or answers immediately.
- Displays error messages if commands are incorrect.
- Interactive mode helps you to learn Python, test new ideas and check grammar or *syntax*.
- When typing quite complex commands into the **Console** like those on the previous page, it's quite easy to make a mistake. Then you have to type the whole line into the **Console** again.
- As discussed in Chapter 5, when using the **Editor** in script mode, it's very easy to correct any mistakes and save the program, without re-typing the whole command, unlike interactive mode in the **Console**.

Code Completion

This feature suggests possible words when you start typing in the **Console**. For example, if you've just used a raw-input( ) command, this will be displayed if you start typing the same command again. Select the suggestion to save typing time.

Code Completion must be switched **On** in **Settings** in the **Script Library** as discussed on page 22. Tap the gear icon to open **Settings**.

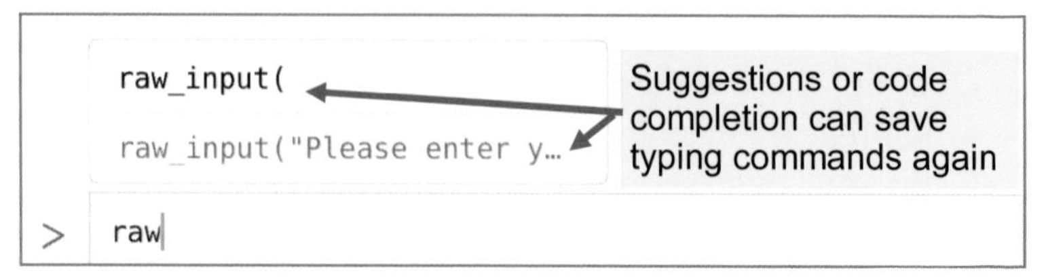

5

Using the Editor

Introduction

Chapters 3 and 4 showed how you can use the Pythonista **Console** to enter one-line commands in *interactive* or *immediate* mode. These are useful to test your ideas and to learn what works in Python. The built-in *debugging* feature helps you to find and correct any mistakes.

In contrast to the **Console**, the **Editor** is used for creating *programs* or *scripts* which can contain a large number of instructions or *statements*. Obviously you wouldn't wish to type in a large program every time you wanted to *run* or *execute* it. So the **Editor** allows you to save the program as a *.py file* on the Internal Storage of the tablet. The commands in the Python language are the same on different *platforms*, i.e. types of computer. This allows Python files to be transferred between computers or a copy given to someone else, as discussed in Chapter 10. This chapter shows how to:

- *Launch* or open the Pythonista Editor.
- *Write* a script consisting of several lines of code.
- *Save* the program as a .py file.
- *Open* or retrieve the file from the Internal Storage.
- *Run* or *execute* the program.
- *Edit* or correct the program to *debug* any errors.

Launching the Editor

Tap the icon shown on the right on the **Apps** screen, then swipe in from the left or right to display the **Scripts Library** as shown on page 21.

From the menu at the bottom left of the **Scripts Library** tap the icon shown on the right and on page 22. From the menu which appears, tap **Empty Script** to open a blank **Editor** screen, shown partly below.

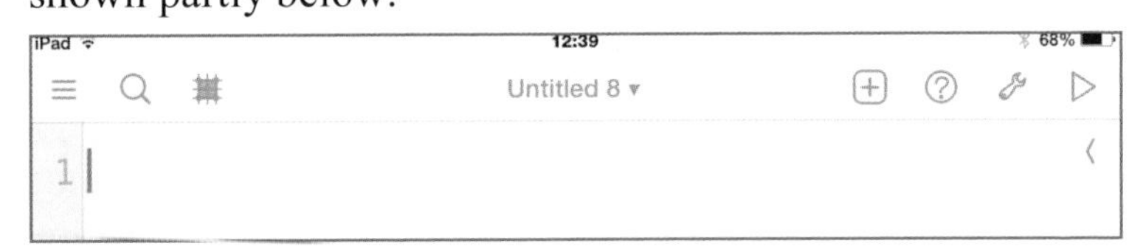

Tap anywhere on the screen to display the on-screen keyboard. You are now ready to start entering and saving your first program, as discussed on the next few pages.

Some of the most important icons at the top of the **Editor** screen shown above are as follows:

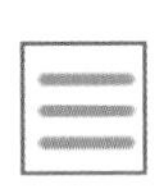

Switch or "toggle" between the **Editor** and the **Scripts Library**. For example, to go to the **Scripts Library** to change **some Settings**.

Tap here to save a script as a file with a suitable name. In Pythonista the .py file name extension is added automatically.

Display the Pythonista **Documentation** as shown on page 20.

Run or execute the script which is currently open in the **Editor**.

Line Numbers

The left of the screenshot shown on page 44, shows line number 1, instead of the ***command prompt*** shown on the right, which is used in the **Console**. When you tap to start entering some code, the on-screen keyboard appears, as shown on page 23.

>>>

After you start entering lines of code and pressing the **return** key, the lines are automatically numbered, as shown below. Don't worry if you don't yet understand the meaning of this particular piece of code — it will be explained shortly.

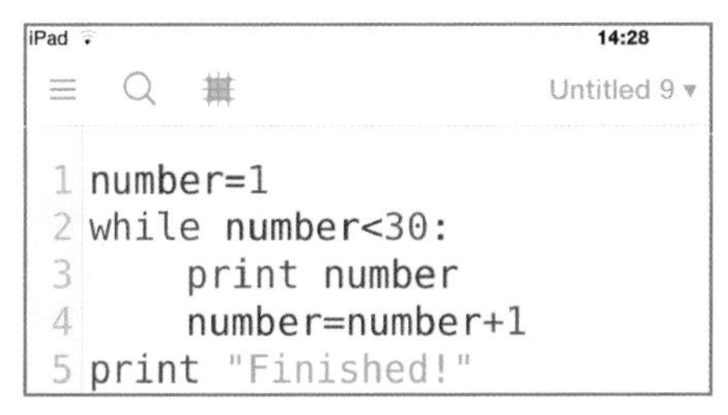

Reserved Words

The screenshot above is just meant to show that a program is a list of code or instructions. Pythonista automatically puts words like while and print in a different colour. These are two of the 31 ***keywords*** or ***reserved words*** shown below, which are used in the Python 2.7 language.

and	del	from	not	while
as	elif	global	or	with
assert	else	if	pass	yield
break	except	import	print	
class	exec	in	raise	
continue	finally	is	return	
def	for	lambda	try	

Python 2.7 reserved words

Settings

After you start using the **Editor** you might want to change some of the **Settings** such as the **Editor Font**, **Editor Font Size** or the **Color Theme**. Switch to the **Scripts Library**, as discussed on page 44 and tap the gear icon at the bottom of the screen, as discussed on page 22, to display the **Settings** menu shown below.

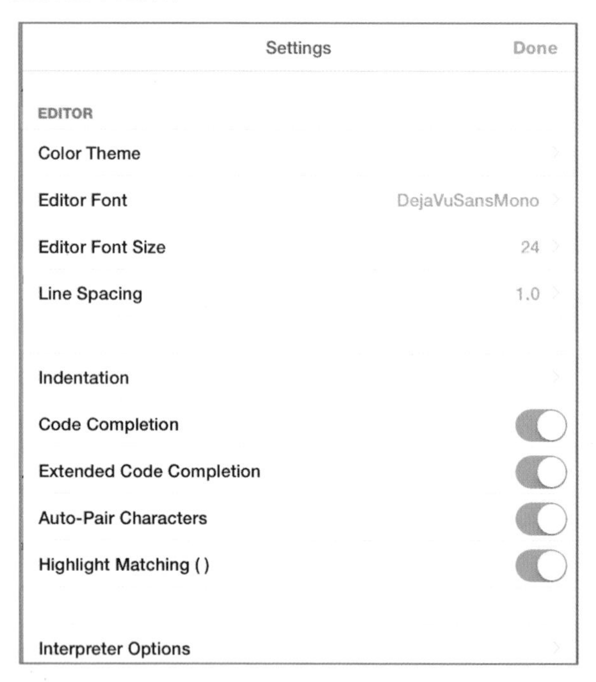

Tap anywhere on a line to change a setting such as **Color Theme**. Select the setting you want then tap **Done**. One of the alternative **Color Themes** for a script is shown here on the right.

```
1 number=1
2 while number<30:
3     print number
4     number=number+1
5 print "Finished!"
```

Indentation

Notice that lines 3 and 4 on the right are indented by a number of spaces (usually 4) and this has a special purpose. The indentation is inserted automatically when you tap **return** after a colon (:) in a while statement. The amount of **Indentation** can be changed as shown below, in the **Settings** in the **Script Library**, as discussed on page 46.

```
1 number=1
2 while number<30:
3     print number
4     number=number+1
5 print "Finished!"
```

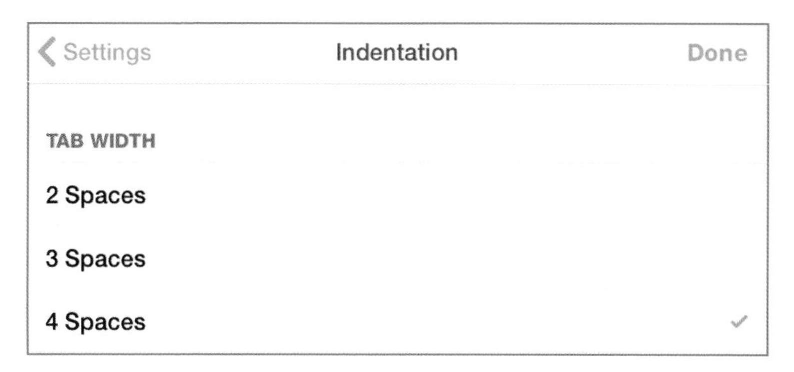

Tap to tick the required amount of **Indentation** as shown above then tap **Done**.

As discussed in detail later, the indented lines, i.e. lines 3 and 4 at the top of this page, are repeated in a ***loop*** as mentioned on page 13. After the indented lines have been repeated the required number of times, program execution continues downwards to the next line which is not indented, i.e. line 5 in the small program at the top of this page.

A colon followed by a block of indented lines is also used after a for statement and an if statement, as discussed later.

Coding and Saving a Program

This section shows how to enter and save a simple program or script. Open a new, blank script in the Pythonista Editor, as described on page 44.

Type the following script into the **Editor**, although you may wish to type your own name instead of "John Brown". Press **return** at the end of every line. The "less than" (<) and "greater than" (>) keys appear in the top row on the on -screen keyboard.

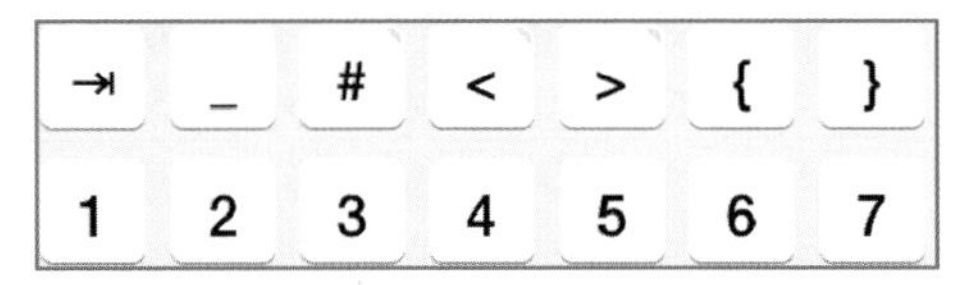

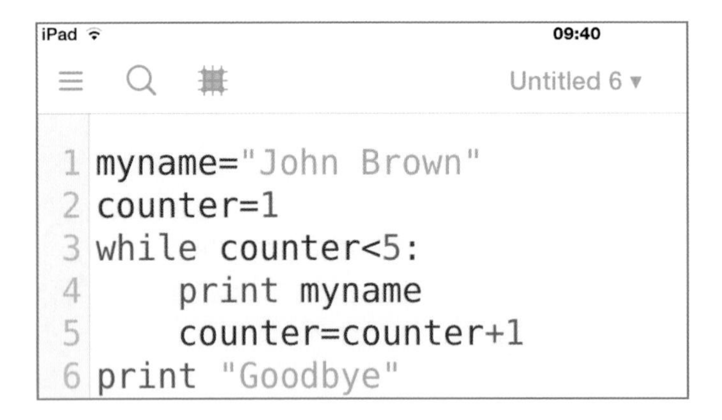

When entering lines after an indented block, i.e. that are not to be indented, such as line 6 above, you need to physically remove the automatic indent using the backspace key shown on the right.

Notice how Python displays keywords such as while and your own words such as "Goodbye" in different colours.

Hints for Avoiding Coding Errors

- Make sure all of the quotes and the brackets (or ***parentheses***) are present.
- If you prefer you can use single speech marks as in 'John Brown' rather than double, as in "John Brown".
- The colon (**:**) must be present at the end of line 3.
- Lines 4 and 5 must be indented by the *same amount* (normally 4 spaces).
- Lines 1, 2, 3 and 6 should *not* be indented at all.

If in doubt about the correct ***syntax*** or ***grammar*** of a line you can quickly test various alternatives in ***interactive mode*** in the **Console**, as discussed in Chapter 3.

Saving a Program

Until you save a program as a .py file, a default name, such as Untitled 6 appears on the **Editor** screen, as shown on the program listing on page 48. Tap over the Untitled name then tap the pencil icon shown on the right and below.

Untitled 6 ▾

Then type the required file name, such as myname, in this example, into the bar as show below. There's no need to add the .py extension in Pythonista — it's done automatically. Press **Done** to complete saving the file.

Displaying a New File in the Script Library

After you've tapped **Done** to save the file you can see it listed in the **Scripts Library**. From the **Editor**, tap the icon shown on the right to switch to the **Scripts Library**, as shown below. You may need to scroll up or down in the left-hand panel to see your particular file amongst the many sample scripts provided in Pythonista.

Executing a Program

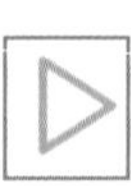

After you've viewed the script you can return to the main **Editor** screen by tapping the icon shown on the right. In the **Editor** you can *execute* the program by tapping the **Run** icon shown on the right and on the menu bar on page 44. The output from running the little program shown on page 48 is displayed below.

John Brown
John Brown
John Brown
John Brown
Goodbye

Managing Your Scripts

With the **Scripts Library** displayed as discussed on page 50 , tap, **Edit** shown below at the top of the screen near the centre.

Edit

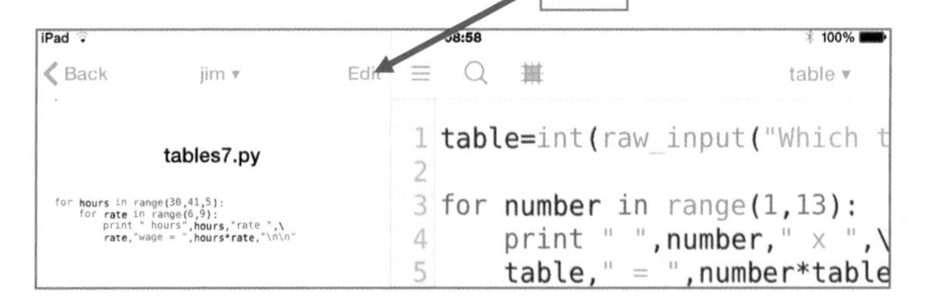

Please note that the **Edit** button shown above in the **Scripts Library** is used for ***managing files***, i.e. ***deleting***, ***moving***, etc. This not to be confused with the main **Editor** screen described on page 44 in which you ***type***, ***edit*** and ***save*** the ***text*** of the ***script*** before ***running*** it.

After tapping **Edit** as shown on the previous page, the **Delete...**, **Move...** and **New Folder...** options appear at the bottom of the left-hand panel in the **Script Library**, as shown below and on page 50.

You can select one or more scripts to be deleted or moved by tapping the name of the script or its thumbnail. As mentioned on page 22, you can switch between **List** view or **Thumbnail Previews** after tapping the icon shown on the right.

With one or more scripts selected you can then **Delete** them or **Move** them to another folder. In the example below, some of the scripts used in this book are shown in a **New Folder** called **Jim** which has been created. Tap **Done** shown below when you've finished managing your files.

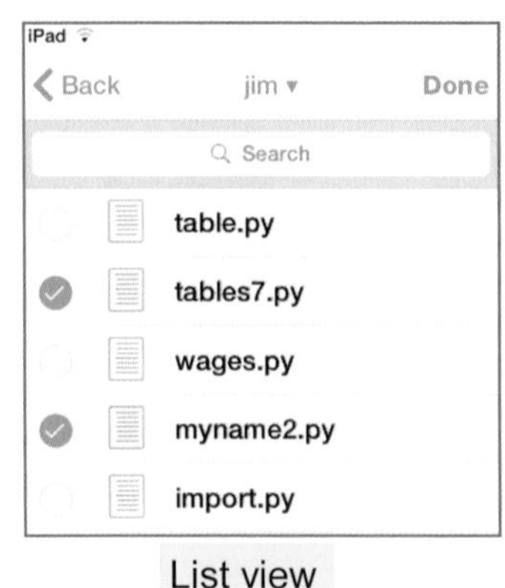

List view

Thumbnail Previews

Adding a Comment

A *Comment* is just a note in a program intended to help someone else to understand the program.

A comment always starts with the # sign and is *ignored* by the computer, e.g.

```
# This prints your name several times
```

Improving the Readability of the Output

Adding Blank Lines Using print

To make the output more readable you can add the word print on its own on a new line, as shown in line 6 below. This "prints" blank lines, as shown on the next page.

Adding Spaces to the Output

You could add some spaces, within speech marks, to the text in the print statements as shown in lines 5 and 9 below. You must put a comma in line 5, before myname.

Save and run this modified program, The new output with more spacing than the original on page 48 is shown on the next page .

```
John Brown

John Brown

John Brown

John Brown

Goodbye
```

Exercise

1. Open the **myname** program shown on page 53.
2. Edit the program to print the name of a pet.
3. Change the script to print the name 8 times.
4. Enter a different message instead of "Goodbye".
5. Use print to print 2 blank lines after every line of output.
6. Experiment with a different number of spaces in quotes before the comma in line 5.
7. Save the program with a new name such as **mypet.py**. (Pythonista adds the .py file extension automatically).
8. Run the program and check for any errors.
9. If necessary, debug the program and save it again.

6

for Loops and Lists

Introduction

Loops are used to harness the power of computers, including tablets and smartphones, to repeat operations at great speed. The while loop, discussed in Chapter 7, keeps repeating a block of commands while something is True.

The for loop is used to repeat a block of commands a specified number of times. This number may be specified directly in the for command. Alternatively, as discussed shortly, the for loop may pass over a ***list*** containing a fixed number of objects.

A simple example would be to display your name on the screen 3 times.

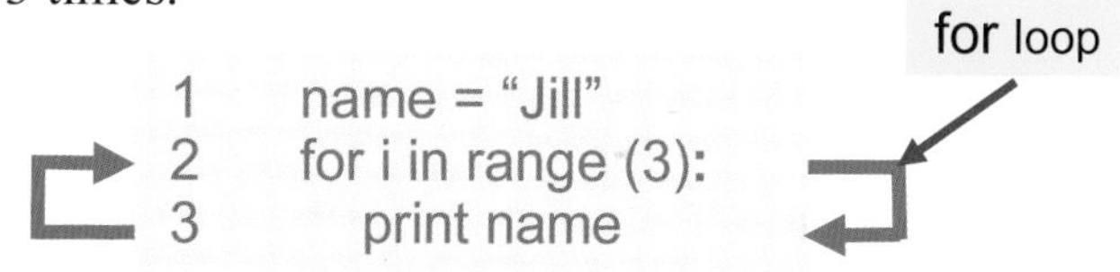

Please note in the above example:

- The colon (:) is essential.
- Lines under the for statement which are to be repeated in the loop must be ***indented*** by the same amount, usually 4 spaces.
- Each journey round the loop is known as a ***pass*** or an ***iteration***.
- It's standard practice to use i and j as variables in a loop.

When you run the program on page 55, the output on the screen is as follows:

```
Jill
Jill
Jill
```

You can improve the output by adding some spaces before the name and by inserting the ***new line*** characters \n, as in:

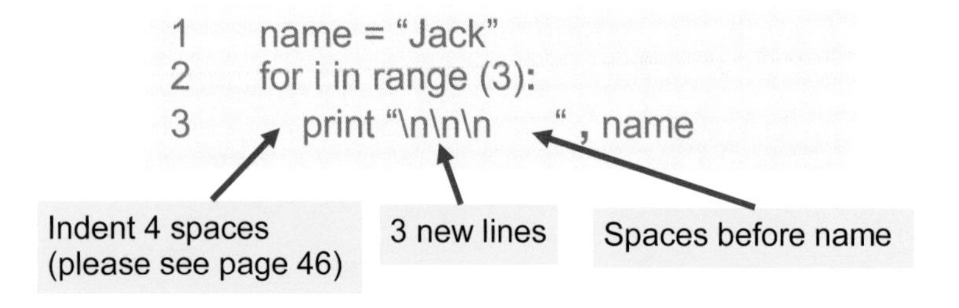

When you run the modified program, the screen display is as shown below:

```
            Jack


            Jack


            Jack
```

Please also note that a comma is needed between separate items in a print statement.

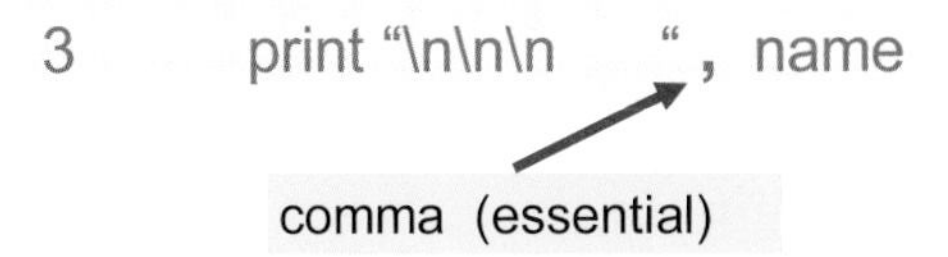

Printing Text on the Same Line

In the previous example, the names were printed underneath one another. In the top example on the previous page, on each pass through the loop "Jill" is printed on a new line. In the bottom example on page 56, the new line characters \n are inserted to give extra spacing. If you want to display data from the loop on the same line, you need to remove the \n characters and insert a comma at the very end of the line, as shown below.

print on the same line

```
3        print "     ", name,
```

The effect of this comma at the end of the line is to display the names on the same line as shown below. To increase or decrease the separation between the names, adjust the number of spaces between the quotes shown above.

```
Jack      Jack      Jack
```

Exercise: Write a program to display your own name 10 times, down the screen. Experiment with spacing. Save the program then edit it to display your name ***along*** the screen, horizontally.

Printing Numbers in a Range

The numbers in the range can be displayed using the code shown below:

```
1    for i in range (10):
2        print i, "     ",
```

The output from the two lines of code above is shown below. Please note above that, unless otherwise stated, the for loop always starts from 0 and finishes at 1 below the number in the range.

```
0    1    2    3    4    5    6    7    8    9
```

We can also specify a ***starting value*** for the loop, as in:

```
1    for i in range (1,11):
```

This displays the following:

```
1    2    3    4    5    6    7    8    9   10
```

The comma at the end of line 2 above ensures that the numbers are displayed ***across*** not ***down*** the screen.

To specify a step up or down in the series of numbers displayed, enter a third number in the brackets in the for loop:

```
1    for i in range (1, 20, 3):
2        print i, "    ",
```

The output is as follows:

```
1   4   7   10   13   16   19
```

Similarly, you could step down by inserting a negative number in the bracket:

```
1    for i in range (30, 0, -3):
2        print i, "    ",
```

As shown below, although a finishing value of 0 was specified, the screen display stops at 3.

```
30   27   24   21   18   15   12   9   6   3
```

This is because the loop is terminated as soon as variable i contains the number 0 and before displaying 0 on the screen. To ensure that 0 is displayed on the screen, as shown below, change line 1 as shown below.

```
1    for i in range (30, -1 , -3):
2        print i, "    ",
```

Displaying a Multiplication Table

The small program below allows the user to choose a multiplication table to display on the screen.

```
1 table=int(raw_input("Which table?  "))
2
3 for number in range(1,13):
4     print "  ",number," x ",\
5     table," = ",number*table
```

In line 1 raw_input stops the execution of the program until the user types some data and taps **return**. "Which table? " is a prompt asking the user to enter a multiplication table, such as 9. This is assigned to a variable called table. In line 1 int ensures that the data entered is treated by Python as a ***number*** and not a ***string***.

The for loop needs to work out the table for the numbers 1 to 12, so it's necessary to enter 13 as the top of the range, as discussed on page 58.

In lines 4 and 5 above, the text inside of the quotes is displayed literally on the screen. number is the ***pass*** or ***iteration*** around the loop ranging from 1 to 12. Note the commas between items in lines 4 and 5.

In line 4 the backslash character \ allows a long statement to be split between two lines.

```
Which table?  9
   1 x 9 =  9
   2 x 9 = 18
   3 x 9 = 27
   4 x 9 = 36
   5 x 9 = 45
   6 x 9 = 54
   7 x 9 = 63
   8 x 9 = 72  etc.,
```

The program on the previous page can be represented by a flowchart, as shown below.

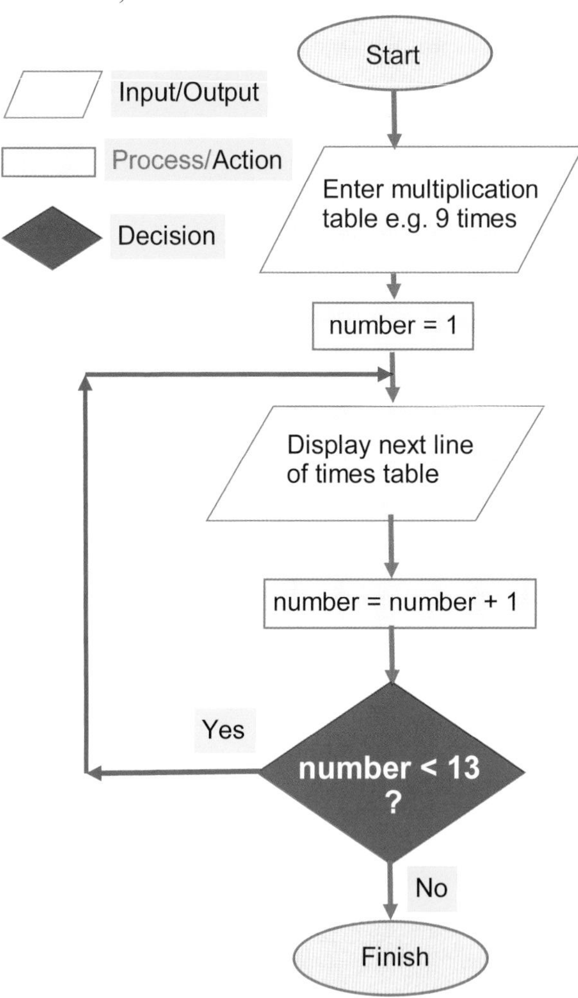

A Flowchart to Display Any Multiplication Table

Nested Loops

Sometimes it's necessary to have ***nested*** loops or loops within loops. This would occur if you have two ranges for two variables. For example, to calculate a weeks wages for different hours worked and different rates of pay.

In the example below, we have a range of hours worked from 30 to 40 in steps of 5, i.e. 30, 35 and 40. As discussed on page 58, to use a value of 40 we need to specify 41 in the hours range. Similarly, the rates of pay are 6, 7 and 8, so we need to specify 9 pounds as our upper hourly pay rate.

```
1 for hours in range(30,41,5):
2     for rate in range(6,9):
3         print " hours",hours,"rate ",\
4         rate,"wage = ",hours*rate,"\n\n"
```

As shown above, there is an outer loop for the hours worked and an inner loop for the rate of pay per hour. The inner loop is indented by four spaces, as is the block of commands shown in lines 3 and 4 above.

First the outer loop is executed with 30 assigned to the hours. The inner loop is repeated 3 times for the range of pay rates 6, 7 and 8 pounds per hour, as shown at the top of the next page.

Next the outer loop is executed again with the hours set at 35 and the whole of the inner loop executed with a total of 3 passes or iterations. Finally the inner loop is executed 3 times with the outer loop set at 40 hours.

```
hours 30  rate 6  wage = 180
hours 30  rate 7  wage = 210
hours 30  rate 8  wage = 240
hours 35  rate 6  wage = 210
hours 35  rate 7  wage = 245
hours 35  rate 8  wage = 280
hours 40  rate 6  wage = 240
hours 40  rate 7  wage = 280
hours 40  rate 8  wage = 320
```

Please note in the program on page 62:

- The upper value in the range must be set at 1 higher than the actual value required; so to have an upper value of 40 hours we need to specify 41.
- for statements must end with a colon (:).
- Each statement in the ***block*** of text under the for statement must be indented by the same amount.
- It's usual to indent each line of a block of text by four spaces as discussed in Chapter 5.
- The backslash \ character at the end of line 3 on page 62 allows a long statement to be split between two lines.

Exercise: Copy, save and run the wages program on page 62. Then edit the program to change the hours worked and the rates of pay to your own values.

Storing Data in Lists

So far we have assigned individual pieces of data to one variable, such as:

catsName = “Serina”

A ***list*** allows you to use a single variable name to assign multiple items of data, as shown below:

ourCats=[“Serina”, “Coco”, “Crisp”, “Halebop”,“Meadow”]

Please note that the list is enclosed by ***square brackets***, which appear on the top row of the on-screen keyboard (as shown on pagc 4).

A list can also include ***numbers*** as well as the ***strings*** shown above, or a mixture of strings and numbers.

So for example we could have the following short program listing sales figures for representatives in the UK.

```
1 salesNorth=["Smith",23,"Jones",31]
2
3 salesSouth=["Scot",38,"Brown",17]
4
5 salesUK=salesNorth + salesSouth
6
7 print salesUK
```

Line 5 links or ***concatenates*** the two lists to produce the single list, salesUK printed by line 7 as shown below.

[‘Smith’, 23, ‘Jones’, 31, ‘Scot’ 38, ‘Brown’, 17]

Each of the individual items in a list is ***indexed***, starting with [0]. So, the first four items in the ourCats list on the previous page are :

ourCats [0] = "Serina"   ourCats [1] = "Coco"

ourCats [2] = "Crisp"   ourCats [3] = "Halebop"

Please note that in a list of 5 items, since the first item has an index of 0 the fifth item has an index of 4. The full list can be displayed using a for loop, as shown below:

```
1 print"\n\n\n"
2
3 ourCats = ["Serina","Coco",\
4 "Crisp","Halebop","Meadow"]
5 for cat in ourCats:
6     print "    ",cat,
```

In the above example, print "\n\n\n" is used to put some blank lines above the output on the screen.

The backslash \ character at the end of line 3 above allows a long program statement to be split between two lines.

Iterating Over a List Using a for Loop

In the statement below, cat is a variable name made up for use in the for loop. ourCats is the name of the list.

```
5    for cat in ourCats:
```

The indented block (just the print statement in line 6 in this example) produces the screen output shown below.

```
Serina   Coco   Crisp   Halebop   Meadow
```

In the output at the bottom of the previous page, space at the top of the screen is created by print "\n\n\n" in line 1. Space between each name is achieved using "   " in line 6. The comma at the end of line 6 displays the output horizontally, across the screen. Without the comma, by default, the cats' names would be printed underneath one another.

Changing an Item in a List

You might need to change a piece of data in a list. For example, in the ourCats list on the previous page, we might want to replace Coco with Claud.

In the previous list:

```
ourCats[1] = "Coco"
```

To change Coco to Claud we would add this extra line:

```
ourCats[1] = "Claud"
```

Printing the Last Item in a List

The last item in a list is always indexed [-1], so to print the last name in the ourCats list we can add the line:

```
print ourCats[-1]
```

The modified lines to be added to the program on page 65 are shown below.

```
 8 ourCats[1]="Claud"
 9 print "\n\n\n"
10 for cat in ourCats:
11     print "    ",cat,
12
13 print "\n\n\n   ", ourCats[-1]
```

When you run the modified program, the output is as follows:

```
Serina   Coco    Crisp   Halebop   Meadow

Serina   Claud   Crisp   Halebop   Meadow

Meadow
```

The first line above was the original ourCats list. In line 2 above Coco has been replaced by Claud. Line 3 above shows the output from the item indexed [-1], i.e. Meadow, the last cat in the list.

Adding an Item to a List

To add another name to the cats list add something like :

```
ourCats.append ("Charlie")
```

Removing an Item from a List

To remove the third item from the list, add the statement:

```
del ourCats[2]
```

(Remembering that the first item is ourCats[0].)

The append and del statements can be added to the end of the cats program as shown on the next page.

```
15 ourCats.append("Charlie")
16
17 del ourCats[2]
18
19 for cat in ourCats:
20     print "  ",cat,
```

Line 15 above adds Charlie to the end of the list as shown below. Line 17 deletes item 3, indexed as ourCats[2], i.e., Crisp, from the list. The for loop at line 19 produces the following modified output:

```
Serina   Claud   Halebop   Meadow  Charlie
```

Tuples

A ***tuple*** is similar to a list but the tuple can't be modified, unlike the ourCats list just described. Tuples are used for items which don't change, such as the months of the year, star signs, dates of birth, etc. A tuple is enclosed in round brackets ( ) rather than the square brackets [ ] used in lists.

Exercise: Copy, save and run the program shown on page 65. If necessary, debug the program and save it again. Then edit or rewrite a program to create and display a list of your own, with ten items of data and your own variable names instead of ourCats and cat. Replace, delete and append items using the methods described on the previous pages.

7

while and if Statements

Introduction

One of the great advantages of computers, including hand-held devices such as the iPad and the iPhone, is their ability to rapidly repeat a task a large number of times. So, for example, it's just as easy to display the numbers from 1 to a 1000 as it is to display the numbers from 1 to 5.

Shown below is a small while loop which prints the numbers from 1 to 5 inclusive.

```
1 # numbers 1 to 5
2 number=1
3 while number<=5:
4     print number
5     number=number+1
6 print "The loop is finished"
```

After saving and running the program, the output on the screen is as follows.

```
1
2
3
4
5
```

To change the program to print the first 1000 numbers (or even 10,000 or 1,000,000) numbers it's simply a case of changing line 3 shown below. (Line 1 is only a comment for information purposes.)

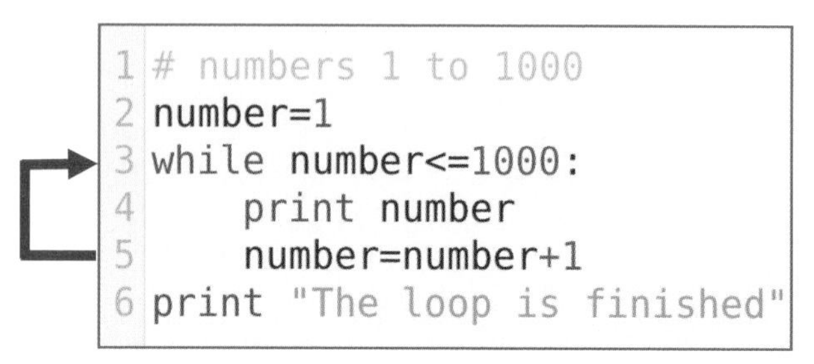

```
1 # numbers 1 to 1000
2 number=1
3 while number<=1000:
4     print number
5     number=number+1
6 print "The loop is finished"
```

When you save and run this modified program, as described on pages 49 to 51, the new screen output is as shown in the small sample below. The computer displays all 1000 numbers almost instantly.

```
 996
 997
 998
 999
1000
 The loop is finished
```

This small example is just intended to illustrate the awesome power of a computer – imagine writing out the first 1000 numbers by hand!

Exercise: Copy, save and run the above program. Then change 1000 in line 3 to 10,000 and save and run the program, as discussed on page 49 and 51.

The Program Statements in English

1. This line is a ***comment***, ignored by the computer
2. Assign an initial value of 1 to the variable store we have called number.
3. While this statement is True, execute the indented lines below it. If not branch to the next line which is not indented, i.e. line 6.
4. Display on the screen the value or number in the variable store number.
5. Add 1 to the value of variable store number.
6. This line is not indented so it is only executed when the loop has finished i.e. when the while statement is no longer True.

Comments

Comment statements, as mentioned above, are simply notes to help people understand a program listing.

True and False

<= in line 3 means "less than or equal to" as discussed on page 37. The while statement is a condition which is either True or False. So, for example, if number contained, say, 578 or 1000, the condition would be True. If number contained 1001 the condition would be False.

The indented lines 4 and 5 are repeated as long as the while condition is True. When the condition is False the program leaves the loop and jumps to the next line which is not indented, in this case line 6.

Indentation is usually 4 spaces (or 8 spaces for a "nested" loop) or after an if statement as shown on page 73.

The Infinite Loop

All being well, the conditions for a loop are met and the program finishes as discussed on the previous page. However sometimes you might make a mistake which prevents the loop from being completed. For example, if you fail to indent line 5 as shown below.

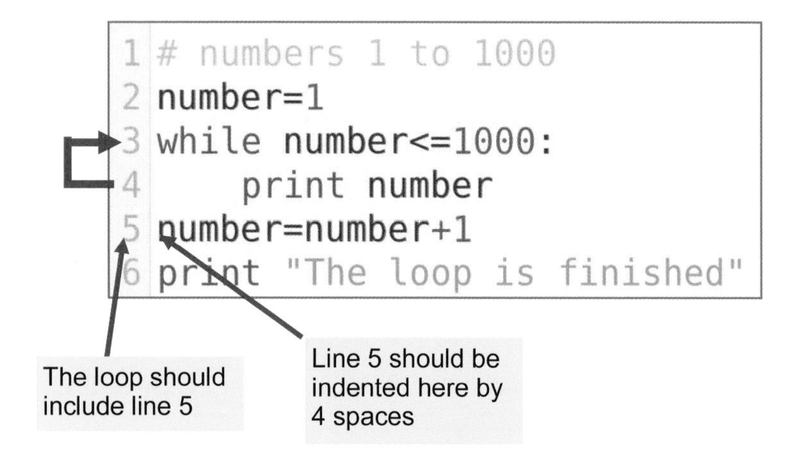

In this case, line 5 will not be repeated in the loop and the store called number will not be increased by 1 with each passage through the loop. The loop will continue but number will remain at 1. So line 3, which says "while number is less than or equal to 1000" will always be True. So the loop will continue, with 1 being repeatedly displayed on the screen until you quit the **Console**, close Pythonista or in the last resort, switch off the iPad or iPhone.

number = number + 1 can be written as number += 1

The if Statement

"If" is used a lot in coding as it is in everyday life., i.e. *if* something is true do one thing, else if it's not true, i.e. false, do something else. In computing these ***conditional expressions*** cause the flow of a program to ***branch*** in different directions. The keywords used in Python for decisions and branching are if, elif and else.

The following program uses if within a while loop. Enter this exactly, making sure you use commas, spaces, etc., as shown in the example. Also the colon in line 4 and extra indentation of another 4 spaces in lines 6, 7 and 8.

The backslash \ at the end of line 7 is used to split the long print statement into two lines.

```
 1
 2 number=1
 3 counter=0
 4 while number <= 84:
 5     if number % 7 == 0:
 6         counter = counter + 1
 7         print "\n", counter,\
 8         " times 7 =", number ," \n"
 9     number=number + 1
10 print "Finished!"
```

After entering this program, save it with a name such as **times7**. (Pythonista adds the .py extension automatically.) As you may have guessed, this program displays the 7 times table. Saving your programs is discussed in more detail on page 49. To run the program, tap the icon shown on the right. Your output should be as shown in the small extract on the next page.

```
 8  times  7 =  56
 9  times  7 =  63
10  times  7 =  70
11  times  7 =  77
12  times  7 =  84
Finished!
```

Correcting Errors –Debugging

If your output is not the same as the extract above, you need to use the **Editor** to correct any mistakes. When a program fails, the error messages on the screen should help. For example, if you miss the second quotes off line 10, as in:

```
10 print "Finished!
```

When you try to run the program, the following red and white **Syntax Error** message is displayed, pointing to the missing quotes in line 10.

```
8          " times 7 =", number ," \n"
9      number=number + 1
10 print "Finished!          SyntaxError: EOL while scanning
```

Correct the error using the **Editor** then save and run the program again. ***Syntax errors*** include mistakes such as missing off a bracket, quotation marks or a spelling mistake in a ***reserved word*** as discussed on page 45.

Understanding the Program

```
 1
 2 number=1
 3 counter=0
 4 while number <= 84:
 5     if number % 7 == 0:
 6         counter = counter + 1
 7         print "\n", counter,\
 8         " times 7 =", number ," \n"
 9     number=number + 1
10 print "Finished!"
```

The meanings of the above lines in English are:

2	Assign 1 to the variable or store called number.
3	Assign 0 to the variable called counter.
4	Repeat the indented lines below as long as the value of number is less than or equal to 84, i.e. True. Otherwise, if line 4 is False go to the next line which is not indented (i.e. line 10).
5	If the remainder equals 0 when number is divided by 7, carry out the indented lines 6, 7 and 8 below. Otherwise go to line 9 and continue in the while loop.
6	Add 1 to the value in the variable store called counter.
7	Display the first part of the next line of the table.
8	Display the rest of the line of the table.
9	Add 1 to the value in the store called number and loop back to line 4.
10	This line is not indented so the program executes it if the while loop is False, i.e. the loop has finished.

The while Loop in More Detail

```
while number <= 84:
```

This loop increases the value in store number from 1 to 84.

Everything which is ***indented*** below while is repeated.

```
if number % 7 == 0:
```

This means "if the ***remainder*** equals 0 when the value in number is divided by 7, execute the indented lines below." (== in Python means ***equals*** or ***the same as*** and corresponds to = in normal arithmetic). This displays the next line of the table as shown in the output on page 74.

If the remainder is not 0, number is not part of the 7 times table and the program carries on to the next number. This continues in the while loop until all the numbers from 1 to 84 have been tested.

Please note that the lines under if are indented by a further 4 spaces in addition to the indentation for the while loop.

Formatting the Output on the Screen

In the print statement on lines 7 and 8 on page 75:

- "   " is used to separate items displayed across the screen.
- Commas must be used between different items in a print statement.
- \ is used to spread a long statement over two lines.
- \n or \n\n, etc., can be used to give vertical spacing of one or more blank lines on the screen.

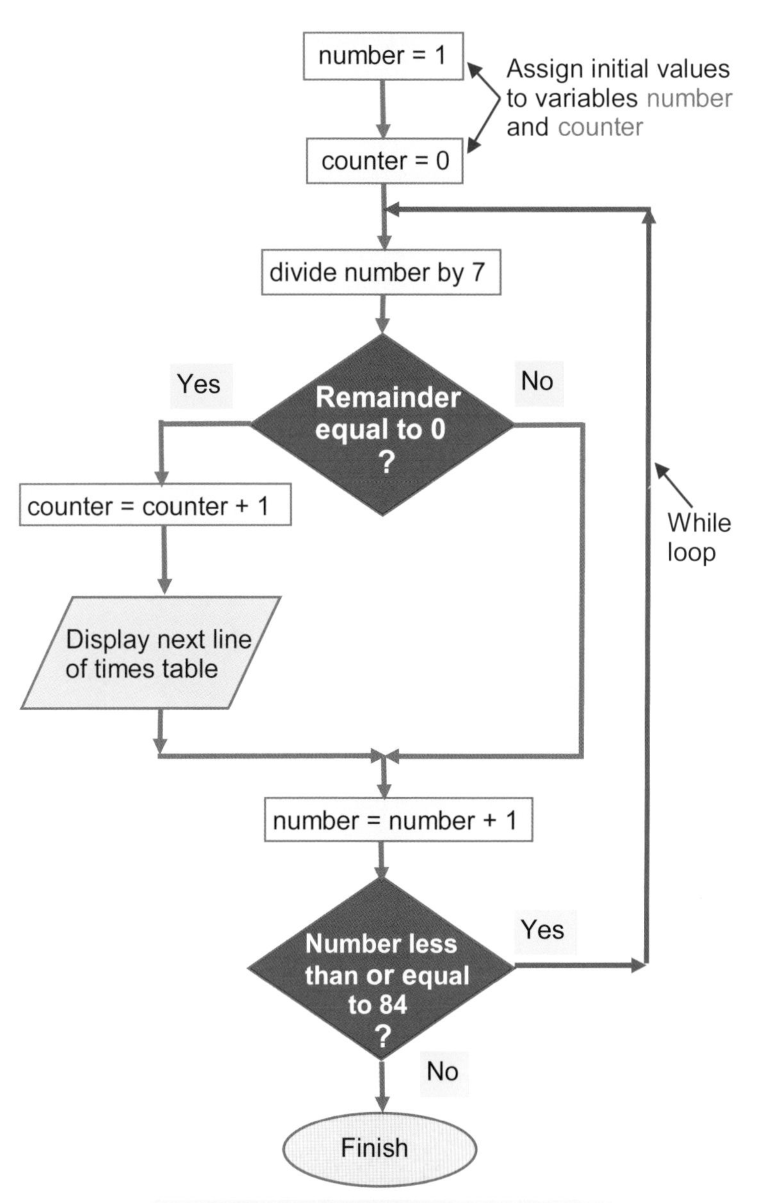

A Flowchart for the Tables Program

Summary: while and if

Although there are simpler ways to "print", i.e. display on the screen, the multiplication tables, the previous example was intended to show the use of the while statement for repetition and the if statement for branching. Both of these statements rely on a condition to be True or False, both end with a colon and use indentation to create a ***block*** of statements which are only executed if the condition is True.

while the condition is True :	if the condition is True :
Keep repeating the execution of this block of indented statements until condition is False.	Execute this block of indented statements then carry on to the next statement below.
Execute this part of the program when the while condition is False.	***Branch*** straight to this line, missing out the indented block when the if condition is False.
while loop	if condition statement

Exercise:

Use the program on page 75 as a template for a program to display the 14 times table up to 252. Replace number and counter with variable names of your own and put your own message in line 10. Save and run the program. If there are any errors, make sure your syntax, i.e. spacing, indentation, commas, quotation marks, colons, etc., are the same as page 75.

8

Further Branching With if, else and elif

Introduction

In the previous chapter, if was used with a single condition statement which was either True or False, as follows:

```
if number % 7 == 0:
    counter = counter + 1
```

The if statement above is testing to see if the remainder is 0 when the contents of the store called number are divided by 7. The above if condition statement has only two possible results.

However, it's not unusual to have more than two possible outcomes to a situation, such as:

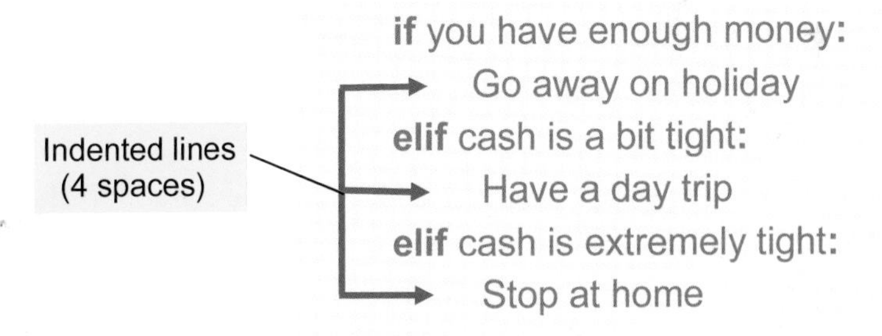

Each of the three condition statements above can be True or False. If a condition is True, the *indented* lines which follow are carried out or executed. Otherwise, if False, the program jumps to the next line which is not indented.

Using the if and else Conditions

In the example below, a password is required, perhaps to be entered before being allowed to use a Web site or enter a building. The person attempting to enter the password would not see this program listing and so would not know the password. In this example the password is python and this has been assigned to a variable called password, shown in line 1 below.

```
1  password="python"
2
3  attempt=raw_input("\n Enter the password: ")
4
5  if attempt == password:
6      print "\n Welcome: Please come in"
7
8  else:
9      print "\n Sorry:Please try again"
10
```

Blank lines are inserted to make the program easier to read. Python uses different colours for keywords, variable names made up by the programmer and text in quotes which is to be displayed on the screen.

Line 3 uses raw_input( ) to ask the user to enter the password. This is assigned to a variable store which has been called attempt. In line 3, 6, and 9, \n causes a new blank line to be inserted, to improve the layout and readability. For the same reason a colon (:) and spaces have been inserted at the end of line 3, as shown below.

("\n Enter the password: ")

```
5 if attempt ==password:
```

This statement tests to see if the password entered by the user and stored in the variable attempt is the same as the actual password in the variable store called password.

== means "is the same as" or "equal to".

It's easy to forget to include the colon (:) on the end of the if and else statements in lines 5 and 8. This will cause a program to fail. The colon is essential as it causes the next line(s) to be indented automatically as required.

When you run this program, if the correct password is entered, the if condition in line 5 is True. So the indented code, line 6, under the if statement, is executed. This displays the following output on the screen.

```
Enter your password: python

Welcome: Please come in
```

If the wrong password is entered, line 5 is not True, so the else statement is executed instead and the following appears on the screen.

```
Enter your password: pyhton

Sorry: Please try again
```

Extending the Program

The listing on page 80 does not allow a user to have another attempt at entering the correct password if their first attempt fails. In practice you are normally allowed several attempts.

So somehow we need to ***repeat*** the process of entering and testing a password. This suggests using a while loop as discussed in Chapter 7 and elsewhere.

What we need is to keep giving the user the opportunity to enter another password until they enter the correct one.

This might be achieved by preceding the attempts by the statement :

```
while attempt != password:
```

The above statement means "while the user's password attempt is ***not the same as*** the actual password, keep repeating the indented lines which follow".

So we could insert the while condition in the listing on page 80, as shown below. Unfortunately the program below fails.

```
1  password="python"
2
3  while attempt != password:
4
5      attempt=raw_input("\n Enter the password: ")
6
7      if attempt == password:
8          print "\n Welcome: Please come in"
9
10     else:
11         print "\n Sorry:Please try again"
```

Reminder:

!= means "not equal to" or "not the same as"

Line 5 on the program on the previous page cannot be executed because no value has yet been assigned to the variable attempt. So the condition True or False cannot be evaluated in line 5 and the error message 'attempt' is not defined appears. This problem can be overcome by assigning an arbitrary initial value to the variable attempt, as shown in line 3 below.

Assigning an Initial Value to a Variable

In order for the while loop (line 5 below) to work, you need an ***initial value*** in the store called attempt.

So line 3 assigns a "dummy" password to the variable attempt, e.g., attempt = "rubbish".

In this case rubbish is used as the dummy but any word would suffice as long as it made the while condition True, i.e. "while the password entered by the user is not the same as the actual password".

```
1  password="python"
2
3  attempt = "rubbish"
4
5  while attempt != password:
6
7      attempt=raw_input("\n Enter the password: ")
8
9      if attempt == password:
10         print "\n Welcome: Please come in"
11
12     else:
13         print "\n Sorry:Please try again"
14
15
16 print "\n \n Have a nice day"
17
```

Running the Extended Program

Now when you run the program on page 83, if the wrong password is entered, the program prints the Sorry message.

```
Enter your password: pyhton

Sorry: Please try again
```

The program then continues in the while loop (lines 5 to 13) on the previous page and allows the user to try again.

When the correct password is entered, the while condition is no longer True. So the program leaves the while loop and prints the Welcome message, followed by the next line which is not indented, i.e. line 16. This displays Have a nice day, as shown below.

```
Enter your password: python

Welcome: Please come in

Have a nice day
```

Exercise:

Copy, save and run the program on page 83, but make up your own password and variable names instead of password and attempt. Debug any errors until it works with both correct and incorrect passwords. Experiment with spaces and \n to improve the display on the screen.

A Bank Account Program

The next program shows how you might use Python to manage your bank account. The following terms are used in everyday life and also as the names of variables in the program.

balance: the amount of money in your account.

credit: a single payment into the account.

debit: a single withdrawal from the account.

The listing for the program is shown below:

```
1
2 print "\n  1. Paying in"
3 print "\n  2. Withdraw cash"
4 print "\n  3. Get advice"
5 print "\n  4. View your balance"
6
7 balance = 100
8 choice = int(raw_input\
9                 ("\n  Enter 1,2,3 or 4\n  "))
10
11 if choice == 1:
12     credit = int(raw_input("\n Enter amount "))
13     balance = balance + credit
14
15 elif choice == 2:
16     debit = int(raw_input("\n Enter amount "))
17     balance = balance - debit
18     if balance < 0:
19         print " \n  You are overdrawn again!"
20
21 elif choice == 3:
22     print "\n Please call in for a chat"
23
24 else:
25     print "\n Your Balance is shown below"
26
27 print "\n\n Current Balance ",balance
28
```

The first four lines of the program shown on the previous page display a menu of options on the screen, as shown below.

1. Paying in

2. Withdraw cash

3. Get advice

4. View your balance

Enter 1, 2, 3, or 4

Using raw_input( ) and int( )

```
8 choice = int(raw_input\("\n Enter 1, 2, 3, or 4 \n"))
```

raw_input( ) (line 8 shown above and on the previous page) causes the program to wait until the user enters something and then presses **return**. Without int( ) and the outer brackets ( ) shown above, the user's input would be treated as a ***string*** not a number. Using int( ) causes the user's input to be treated as a number.

The user enters an option 1, 2, 3 or 4 and taps **return**. The option number is placed in the store called choice.

Depending on the choice, the program branches to one of the lines starting with if, elif, and else shown on page 85.

The use of raw_input( ) to enter strings and integers is discussed in more detail on pages 39-41.

In line 12, int(raw_input (“ “)) is used for the entry of the amount of money to be paid in and credited to the account.

In line 13 the new balance is calculated using:

```
balance = balance + credit
```

After line 13 if choice ==1, none of the elif conditions and the else condition are True. So the program branches to the next line which is not indented, line 27. This displays the current balance, as shown below.

```
Enter amount  65

Current Balance  165
```

Line 17 works out the new balance when the user enters option 2 to withdraw cash, before printing the balance, in a similar way to the credit option described above. The program continues down carrying out the indented lines when the elif or else statements are true. Finally all of the conditions if, elif and else lead to line 27 which prints the balance after whatever choice was selected.

After an if statement you can have as many elif statements as you like. You can only have one else statement. The else statement is optional but may be used to make the code easier to understand.

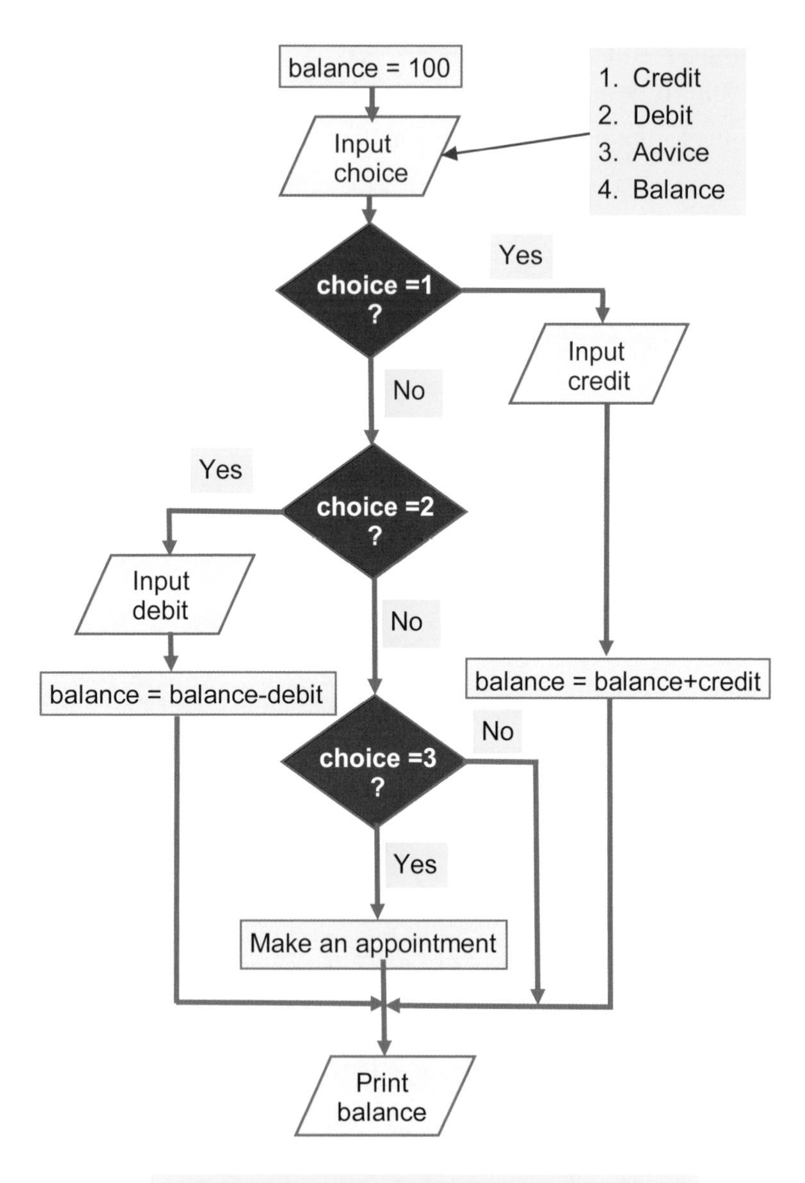

A Flowchart for the Bank Account Program

As shown on the previous page, the diamond-shaped ***decision boxes*** correspond to the if and elif statements on the program on page 85. Each decision box has only two possible results, Yes or No. These correspond to the conditions True or False. As mentioned earlier, a computer, being a ***two-state*** or ***binary system*** can represent the conditions True or False using a system of ***logic gates***.

In the bank program, there is no need for a decision box for the fourth option to print the balance. This corresponds to line 27 in the program on page 85. As this line is not indented, it's executed after all of the conditional statements, if and elif have been tested and executed where necessary. The else statement is not essential since the line which follows will be executed anyway.

Overdrawn?

If you try to take out more than you have in your account, you will be overdrawn and may be charged interest. This can be checked by inserting lines such as:

```
18      if balance < 0:
19          print " \n  You are overdrawn again!"
```

This is inserted after line 17 on the program on page 85. On the flowchart on the opposite page, this would be after the ***action*** box balance = balance-debit. If the condition is True, i.e. the balance is less than 0, the indented print statement shown above is displayed on the screen. The program then continues and "prints" the balance, i.e. displays the balance on the screen.

Exercise

Copy and save the program on listed page 85. Test all of the options 1, 2, 3 and 4 on the menu, using various amounts of money for credits and debits.

Test the overdraft code by selecting option 2 and entering a large cash withdrawal or debit.

If there are any errors, check the code with the version on page 85. Then correct and save the program.

Using Strings with raw_input( )

As discussed on pages 39-41 and page 86, the raw_input( ) function can be used to prompt the user to enter data as strings or numbers. As discussed on page 86, for numerical input, raw_input( ) is preceded by int and enclosed within an outer set of brackets, as shown below.

```
choice = int(raw_input\("\n  Enter 1, 2, 3, or 4 \n"))
```

If you omit int and the outer brackets, in line 8 and 9 on page 85, Python will treat the user's input as ***string data***. So lines 11, 15 and 21 on page 85 will need the choices 1, 2 and 3 to be enclosed in quotes, i.e. "1", "2" and "3".

Exercise

Use the **Editor** to modify your copy of the program shown on page 85.

Change lines 8 and 9 to remove int and the outer brackets.

Change line 11 to: if choice == "1":

Change 2 and 3 in lines 15 and 21 to "2" and "3".

Run the program and debug if necessary.

9

Functions and Modules

Introduction

Anyone who's used a calculator has most certainly used a function. For example, to find the square root of a number, simply enter the number and tap the square root key, usually marked as shown on the right. So if we enter 25 into the calculator and press the square root key, the answer 5 will pop up straightaway. In fact, the process of finding a square root can be quite complex without a calculator or a computer program.

√

For example, to find the square root of 40 we need to find a number which, when multiplied by itself, gives 40 as the answer. One method is to keep guessing, until we get very close, if not exactly, to 40, as shown below.

6.5 x 6.5 = 42.25 6.4 x 6.4 = 40.96
6.35 x 6.35 = 40.3225 6.3 x 6.3 = 39.69

Fortunately this laborious ***iteration*** process is reduced to a single press of the square root key, because routines or algorithms have been written to find square roots. This is very similar to the use of functions in programming.

A ***function*** is a set of program statements representing a frequently used process. The function can be ***called*** and executed by simply entering its name into a program.

Functions in Python

There are many functions available in Python and you can also write your own. Some of the reasons for using functions are :

- As stated earlier, a function is a block of program statements which is used regularly. It would be inefficient if you had to type in the block of statements every time you used them.
- A widely used function may be saved as a file and inserted into lots of different programs.
- You can utilise functions which other people have written.
- Functions allow a long program to be divided up into manageable "chunks", making it easier to understand and develop the program.

Built-in Functions

Some functions are built into Python and can be called by simply typing the function name into a program. You can experiment with functions in ***interactive mode*** by typing the function name into the **Console**.

We've already used some of the built-in functions earlier in this book. These included int( ), raw_input( ) and range( ). The brackets contain the numbers or strings (known as ***parameters*** or ***arguments***) which the function is going to operate on. These are the ***input*** to the function. After the function is executed, any resulting numbers or strings, i.e. ***output***, are ***returned*** to the main program.

Library Functions

Python has many functions stored in a library of ***modules***. A module is a .py file similar to a program file and contains a list of function ***definitions***, usually on the same subject. You can look at the lists of modules and the functions they contain after searching the Internet for "Python Standard Library".

For example, the math module has a long list of mathematical functions in a format such as math.sqrt(x), math.log(x) and math.sin(x).

Unlike the built-in functions, which can be called by simply typing their name into a program, the library functions have to be ***imported*** into a program, as discussed shortly.

(The Python Standard Library also includes a complete listing of the built-in functions mentioned on page 92.)

User-defined Functions

As well as using Python's built-in and library functions that other people have written, you can also write your own, as discussed shortly. For example, a piece of code which is to be used frequently in a long program could be ***defined*** once as a function. Then to use the function throughout the rest of the program it would simply be called by inserting the function name.

Alternatively, if you want to employ a user-defined function in other programs, the function would be saved in a module and then imported into other programs as required.

Defining your own functions is discussed later in this chapter.

Examples of Functions

range( )

The range function can be used with one, two or three ***arguments***, as shown below:

range(x) range (x,y) range(x,y,z)

The range function is often used in a for loop, as shown below.

```
for num in range(12):
    print num,
```

In the above small program, range(12) produces the output:

0, 1, 2, 3, 4, 5, 6, 7, 8, 9, 10, 11

(0 is assumed as the starting value and by default the ***increment*** or ***step*** is 1)

Similarly using range(2,13) in the for loop above, returns:

2, 3, 4, 5, 6, 7, 8, 9, 10, 11, 12

Finally range(2,15,2) displays:

2, 4, 6, 8, 10, 12, 14

The range function with 3 arguments, as shown above in range (2,15, 2), in general takes the form:

```
range(start, finish, step)
```

int( )

A ***floating point number*** (also known as a ***float***), is a number with figures to the right of the decimal point. The int( ) function can be used to convert the floating point number to an ***integer*** (a whole number). You can check this in interactive mode in the **Console**, as shown below.

Passing Parameters

To use a function in a program you would enter its name as a program statement. When the program reaches the line calling the function, the lines of the function are carried out.

In a numeric function like int( ) shown above, you have to supply or ***pass*** numbers, i.e. ***parameters***, between the brackets as input. The function then performs an operation on the parameters and ***returns*** an answer. So in a program, using the int( ) example, 3.7 is ***passed*** to the function and 3 would be ***returned*** to the main program.

As shown on page 97, int( ) is also used to convert numbers, returned as string characters, to integers.

Parameters and Arguments

The words ***parameter*** and ***argument*** are both used to describe the contents within a function's brackets. One definition is that a parameter is a ***variable*** such as x and y in range (x,y), for example, while arguments are the actual numerical values input to a function such as 2 and 10 as in range (2, 10).

raw_input( )

You can use raw_input( ) to give a prompt to the user to type something. The user enters some data before tapping **return**. The program then moves on to the next line.

```
name = raw_input ("Enter your name")
print " Welcome", name
```

In the above example, when you type your name and press **return**, your name is returned from the raw_input( ) function to the main program where the print statement displays it on the screen.

raw_input( ) and Numbers

Anything the user enters in response to a prompt from raw_input( ) is returned to the main program as a ***string***. In the example below, a wage is calculated from the rate of pay (£9 an hour) after the user enters the hours worked.

```
hoursWorked = raw_input ("Enter your hours")
print " Your wage is  ", 9*hoursWorked
```

This gives a ridiculous answer when you enter, e.g. 18 hours, as shown below.

```
Your wage is  181818181818181818
```

The reason this answer is wrong is because raw_input( ) has returned the ***string*** "18", i.e. just the two keyboard characters, "1" and "8", not the mathematical number 18 made up of a 10 and an 8.

So when 18 is passed to hoursWorked as shown on the previous page, instead of multiplying 9x18 (from 9*hoursWorked) and getting 162, the print statement has simply displayed the string "18" a total of 9 times.

Fortunately this error can be corrected using the int( ) function as shown below. This converts the string produced by raw_input( ) to the integer value required by the wages calculation. To input numbers using raw_input( ), enclose the entire raw_input( ) statement in parenthesis (brackets) and precede the statement by int( ), as shown below.

```
hoursWorked = int(raw_input ("Enter your hours"))
print " Your wage is  ", 9*hoursWorked, "pounds"
```

Please note: It's very easy to forget the outer bracket on the extreme right above.

So now the number 18, not the string "18", is returned from raw_input( ), to give the correct wage calculation shown below.

```
Your wage is 162 pounds
```

Exercise

Use the above example to write a program to work out the wages for a different hourly rate of pay. Modify the program to input both hourly rate and hours worked and print out the wage.

The use of raw_input( ) and int( ) to enter strings and integers is also discussed on pages 39-41.

Modules

So far we've looked at a few ***built-in*** functions that can be used directly by simply typing their name into the program you are creating. Many other functions are stored in a library of ***modules***. As stated before, a module is itself a .py file containing the ***definitions*** for a number of functions. So for example, in the math module there are functions such as sqrt(x), factorial(x) and many others.

sqrt(x)

As stated earlier, the square root (sqrt) of a number is another number which, when multiplied by itself, gives the first number. For example:

sqrt(4) = 2   sqrt(9) = 3

factorial(x)

The factorial of a number takes the number and multiplies it by every other number below it, down to 1. For example:

factorial(3) = 3 x 2 x 1 = 6

factorial(4) = 4 x 3 x 2 x1 = 24

Functions saved in modules are identified using the module name followed by a full stop and the function name, such as:

Module name   Function name

math.sqrt(x)

Importing Functions into a Program

To use a function (other than a built-in function) in a program, it must be ***imported*** into the program from the module (.py file) in which it is saved. It's usual to put the import statements at the start of the program listing.

Method 1. Using the from....import Statement

```
1  from math import sqrt
2
3  number = input ("Enter a number")
4
5  answer= sqrt (number)
6
7  print answer
```

Importing function sqrt

Calling function sqrt

Method 2. Using the import Statement

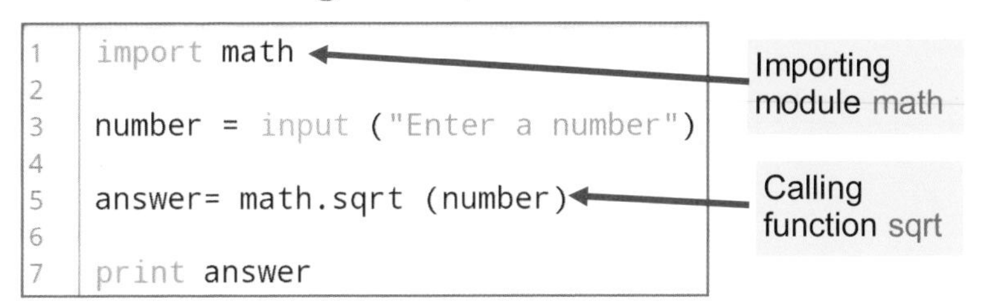

When the above programs are run and the number 20 is entered as the number, both methods yield the same result, for the square root, as shown below.

4.472135955

The round( ) Function

The round(x,y) function discussed on the next page allows you to round a floating point number such as 4.472135955 down to a specified number of places.

input( ) versus int(raw_input( )) in Python 2.7

For simplicity above, the function input( ), rather than int (raw_input( )) has been used. However, Python documentation recommends raw_input( )for general use.

round(x,y)
This function allows you to correct a number, x, to a number of places, y, after the point. So for example,

round(5.391732, 2) would yield 5.39.

As round( ) is a built-in Python function there's no need to import it and so it can be inserted directly into a program, as shown below.

```
1  from math import sqrt
2
3  number = input ("Enter a number")
4
5  answer= sqrt (number)
6
7  print round(answer,2)
```

Correct the number stored in variable answer to two decimal places, i.e. to the right of the decimal point.

When the above modified program is run and the number 20 is entered as input, the answer is as follows:

4.47

Exercise

1. Write programs to find the square root of 30, correct to 3 decimal places, using methods 1 and 2 on page 99.

2. Write a program to import the function math.factorial and use it to calculate and display factorial (20), i.e.

20 x19x18x17x16.........x3x2x1.

How long would this take using pencil and paper?

Random Numbers Using randint( )

There are several functions for generating random numbers within the random module in the Python Library.

For example, randint(1,6) generates random whole numbers from 1 to 6 inclusive. You can experiment with randint( ) using interactive mode in the **Console**. First you have to import the module, random, as shown below.

```
>>> import random
```

Then call the function using the module name, random followed by the function name, randint( ) complete with the required ***arguments***, i.e. values in brackets.

```
>>> random.randint(1,6)
4
```

Output → 4; Input → random.randint(1,6)

So for example, we could simulate throwing a dice and get the result 4, for example, as shown above.

Or we could write a little program using a while loop to simulate throwing the dice 20 times, for example.

```
1 import random
2
3 throws =0
4
5 while throws<20:
6     number=random.randint(1,6)
7     print number," ",
8     throws=throws+1
```

Import random module (line 1); Calling the function (line 6)

The output from the first two runs of the program on page 101 was as follows:

```
5 1 6 2 3 3 5 6 1 5 2 4 6 1 5 4 5 2 6 2

2 3 1 1 5 4 6 2 6 6 1 6 2 6 3 3 4 2 6 2
```

As can be seen above, the algorithm or routine that was devised for the randint( ) function has done a good job in producing two different sets of 20 random numbers between 1 and 6.

Advantages of Functions and Modules

The previous examples show how useful it is to have ready-made functions, either ***built-in*** or available in ***modules***. Many of these functions would be difficult and time-consuming for users to code for themselves. ***Built-in functions*** reduce complex tasks to one simple statement to ***call*** the function using its name. Functions stored within ***modules*** are also called using their name after ***importing*** the module to the program, as described on page 98.

Exercise

Rewrite the program on page 101 to display 10 random numbers between 2 and 9 inclusive. Run the program five times to produce 5 sets of results. (Remember to put the comma at the end of the print statement shown in line 7).

If line 3 was changed to throws=1, try to work out how line 5 would have to change to make sure 10 random numbers were displayed.

Defining Your Own Functions

In this example a simple function, total(x,y), is created to add two numbers. The function is defined and called as shown below. After defining a function it can be used again and again in a program just by entering its name. Or the function could be saved in a module and called after importing from the module into the main program.

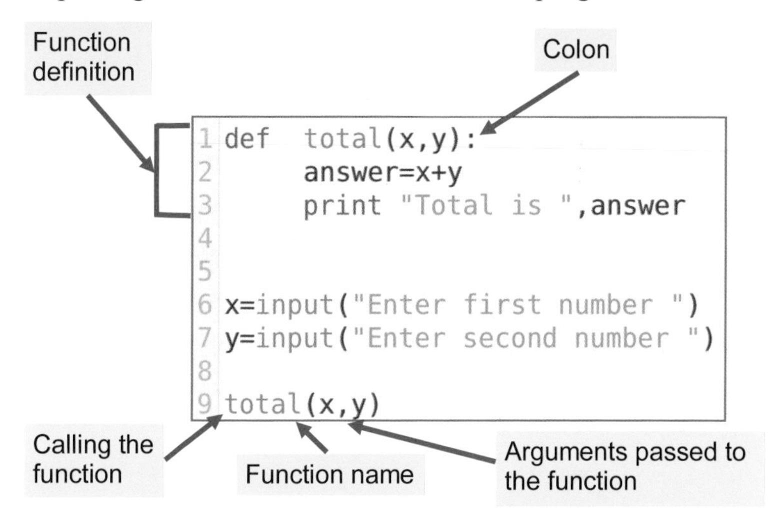

Please note above that the statements which are part of a function definition must be ***indented*** by the same amount, usually four spaces. After the function has been completed, program execution returns to the main program and carries on to the next line after the line which calls the function, i.e. after line 9 in the example above.

Variables declared within a function definition such as answer above cannot be used outside of the function and are known as ***local*** variables. ***Global variables*** are variables declared outside of functions. These can be used everywhere including within functions.

Returning Values from a Function

As stated earlier, variables declared within a function do not apply outside of the function. So in the example on the previous page, answer is a ***local variable*** and its value does not apply in the main program. (Although you could use the variable name answer for a different purpose in the main program with new values assigned to it).

You can use the return statement to send values output from a function back to the main program, as shown below.

```
 1 def  total(x,y):
 2        return x+y
 3
 4
 5
 6 x=input("Enter first number ")
 7 y=input("Enter second number ")
 8
 9 answer=total(x,y)
10
11 print "Total is ",answer
```

In this example the function name is total(x,y) with the arguments x and y, input by the user, being passed to the function. The function adds the two numbers and the return statement sends the result back to the caller, answer above. The ***call*** command has been assigned to the variable store answer in this example.

Exercise

Write a function to find the average of four numbers using def and return as shown above.

10

Working With .py Files Across Platforms

Introduction

Python (.py) files are created on all types of computer platform, as well as the iPads and iPhones discussed in this book. You might do your coding in different places on various machines, such as PCs at school, college or work and a tablet or smartphone at home. Or you might want to send a copy of your latest program to a friend.

You can transfer and run .py files on computers running different operating systems such as iOS (iPads and iPhones), Microsoft Windows (PCs) and also the Android operating system used on many tablets and smartphones.

This book is based on the Pythonista app for iPads and iPhones. Files coded in Pythonista are compatible with the .py files created in Python 2.7 on PC computers. They are also compatible with .py files written using the QPython (but not QPython3) app on Android devices.

Some methods of transferring .py files between the different computer operating systems are listed below.

- ***Upload*** the files from an iPad or iPhone to the "clouds" using Dropbox or Google Drive, etc.
- Use a ***file manager*** to copy the .py files from Dropbox to a PC or an Android device.
- Use ***copy*** and ***paste*** and ***e-mail*** to transfer files from a PC or Android device to an iPad or iPhone.

Sharing Files Using the Clouds

Dropbox and Google Drive are "cloud" storage systems, in which your files, including .py files, are saved on Internet ***server computers***, so they can be accessed from other computers having an Internet connection. When you save a file in the clouds, it is ***synced***, i.e. automatically copied, to all of the other computers on which you have an account with a cloud storage service such as Dropbox or Google Drive, etc.

Although there are other cloud storage systems, such as iCloud and Microsoft OneDrive, I've found Dropbox and Google Drive perform well for the transfer of Python .py files created on different types of computer.

Installing Cloud Storage from the App Store

To share files in the clouds, all computers, including iPads and iPhones, need to have an app such as Dropbox or Google Drive installed, together with a user account. Dropbox and Google Drive are free, although business users can pay more for extra storage space. You can install a copy of Dropbox or Google Drive on an iPad or iPhone from the App Store, as shown below. The icons for Dropbox and Google Drive, shown below in the App Store, are also copied to the Apps screen during the installation.

Installing Cloud Storage from the Web

If you have other computers such as a laptop or a desktop PC you can download Dropbox or Google Drive after opening the websites at:

www.dropbox.com or google.co.uk/drive/download

On a Windows PC machine this will place a Dropbox folder or a Google Drive folder in the left-hand panel of the Windows Explorer/File Explorer, as shown below.

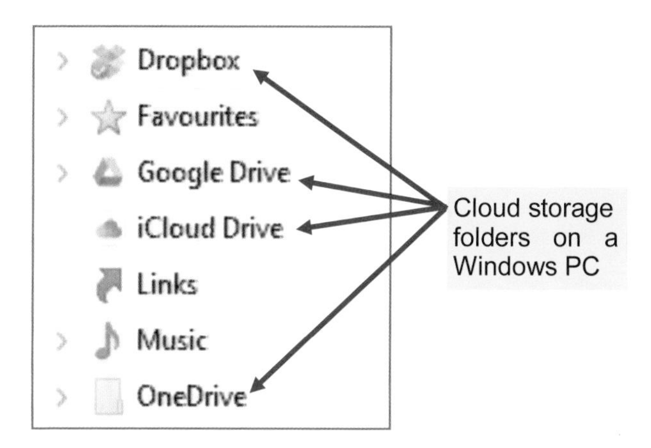

As shown above, this particular Windows PC also had the iCloud Drive and OneDrive cloud storage systems installed, in addition to Dropbox and Google Drive. However, for simplicity, the very popular and well-established Dropbox will be used in the rest of this chapter.

As discussed shortly, you can export your scripts to Dropbox or Google Drive from the Pythonista Editor. Then they can be copied to the Python Editor on a PC or Android device, perhaps so that you can continue working in a different situation. Or a you could send a copy of a script to a friend via a link to your Dropbox folder.

Copying .py Files to Dropbox

This section shows how you can copy or ***upload*** .py files from an iPad or iPhone to Dropbox in the clouds. Then you can copy the files to the Python 2.7 Editor in a PC or the QPython Editor in an Android tablet or smartphone. As discussed earlier, any PC or Android device must have Dropbox installed and a valid Dropbox account.

Open the Pythonista **Editor** on your iPad or iPhone and enter or open the .py script you wish to copy to Dropbox, as shown below. As an example, the small random number program, dice.py, discussed on page 101, will be used.

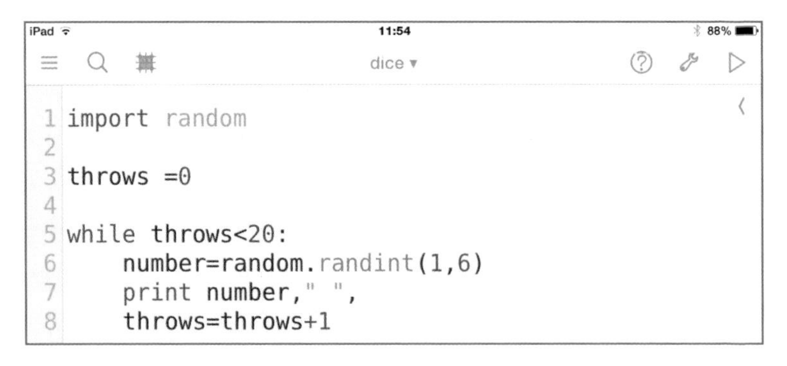

Saving to Dropbox

Tap the small spanner icon shown on the right and in the top right-hand corner of the Pythonista **Editor** above. From the **Actions** menu which appears, tap **Export...** as shown below.

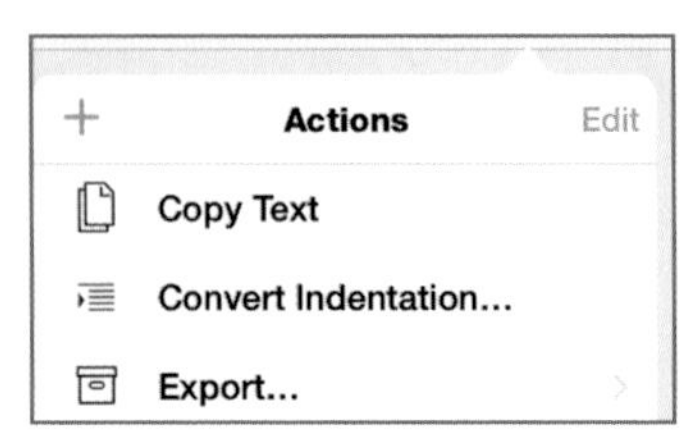

The **Export** window then opens, as shown on the left below, from which you tap **Open in....** This displays some destinations in the clouds to which you can upload the file, as shown on the right below. Tap **Save to Dropbox** as shown on the right and on the right below.

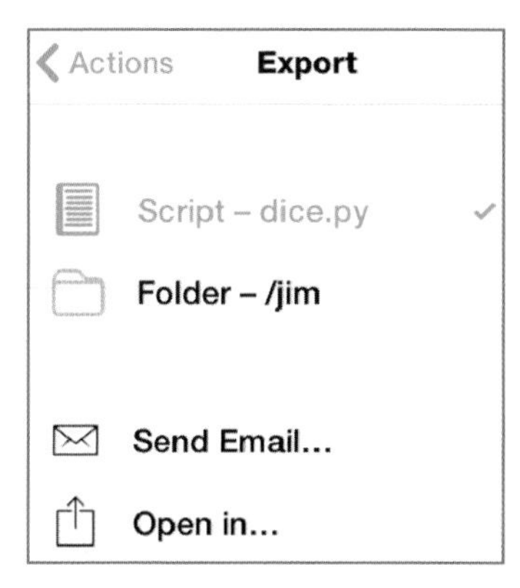

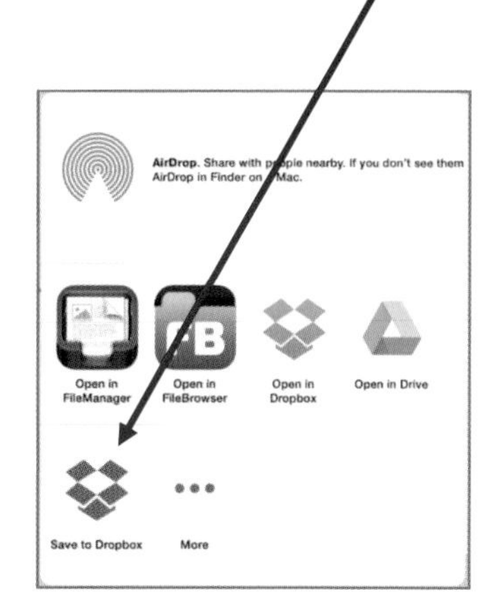

You may be asked to **Sign in** to Dropbox with your **Email** address and **Password** as shown on the right. Or you may need to tap **Create Account** to sign up for a new Dropbox account. Finally tap **Save**, as shown on the right, to copy the file, in this example dice.py, to the main Dropbox folder. To save the file in a personal folder you've created in Dropbox (see page 111), tap **Choose a Different Folder....** Then tap **Save** after selecting your own folder.

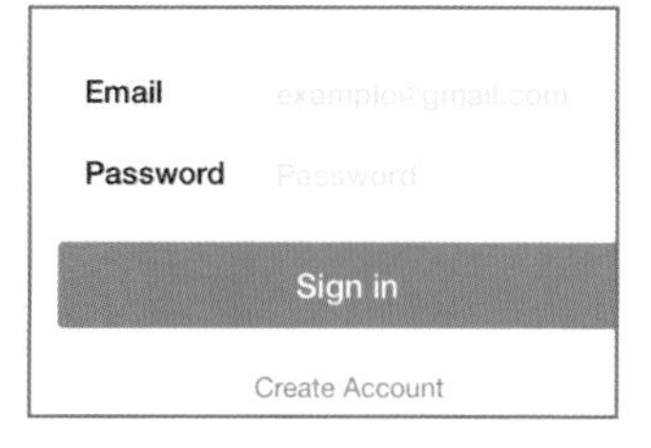

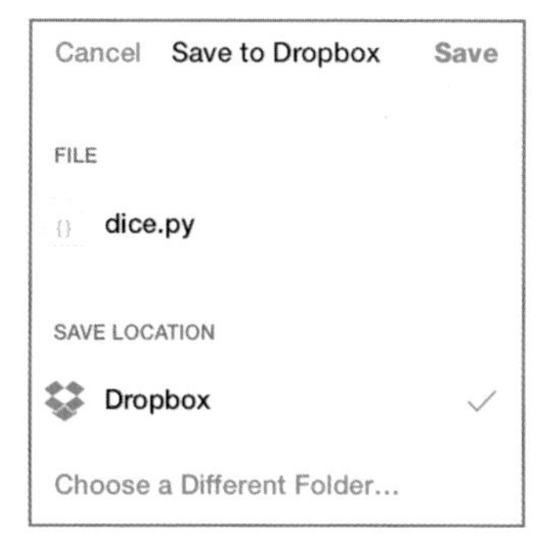

Dropbox on Different Platforms

Once the dice.py file has been saved, the file name can be displayed in Dropbox in your Python folder on different types of computer, as shown below.

The file dice.py in Dropbox on an iPad or iPhone

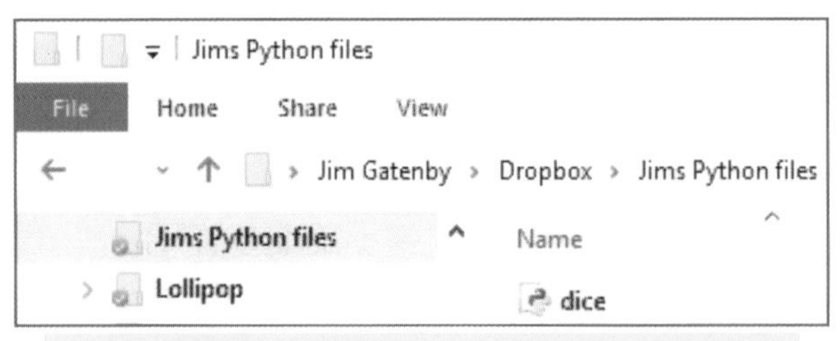

The file dice.py in Dropbox on a Windows PC

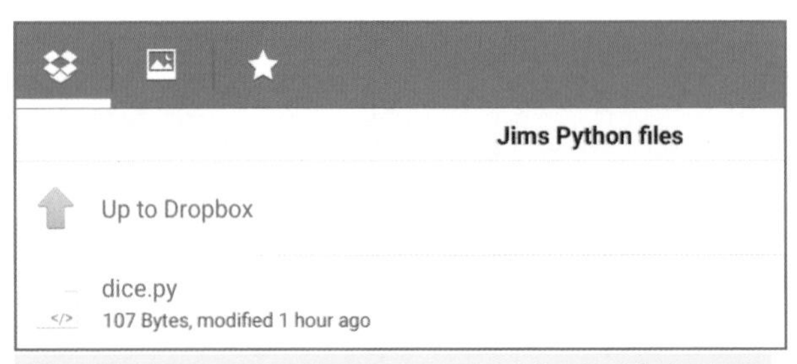

The file dice.py in Dropbox on an Android device

For Python .py files to be compatible across the three platforms shown above, these must all be running the same version of Python. Pythonista and QPython are both based on Python 2.7, also used in PC computers.

To summarise the previous example:

- A file, dice.py, was created in Pythonista on an iPad tablet.
- The file was then uploaded to Dropbox and saved in a folder, **Jims Python files**, which I had created.
- As shown on the previous page, any file you save in Dropbox on one computer is automatically copied or *synced* to your other computers. You must be signed in to Dropbox to view the files and folders.

Creating a Dropbox Folder on an iPad

You can manage your files (including .py files) in Dropbox on an iPad or iPhone, after tapping the icon shown on the right and at the top right of the previous page.

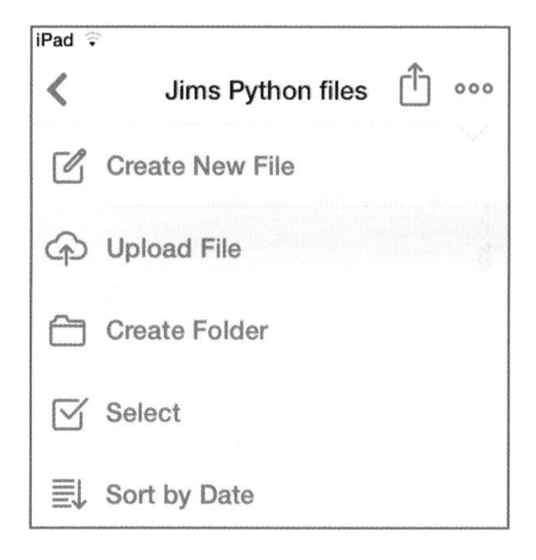

Use **Create Folder** above to make a folder in which to save all of your Pythonista .py files, as discussed earlier. A new folder you create on any of your machines is automatically synced across to all of the other computers on which you have access to Dropbox.

Setting Up a PC to Use Python .py Files

This section assumes you've created a .py file using Pythonista on an iPad or iPhone and copied it to the clouds, using Dropbox, for example. To use a Pythonista .py file on a PC, the PC must be set up with the Python 2.7 interpreter, as discussed below.

Python 2.7 was chosen for the work in this book because Python 3 is relatively new and some modules and functions are still under development. All of the programs in this book have been successfully tested using Python 2.7 with the computer operating systems shown on page 110.

Installing Python 2.7 on a Windows PC

You can install the Python 2.7 app on the PC, after visiting the Web site at:

www.python.org/downloads/.

After tapping the **Download Python 2.7.10** button shown above, select **Run** and follow the instructions on the screen. You can either accept the recommended folder **C:\Python27** for the Python files or select a new folder.

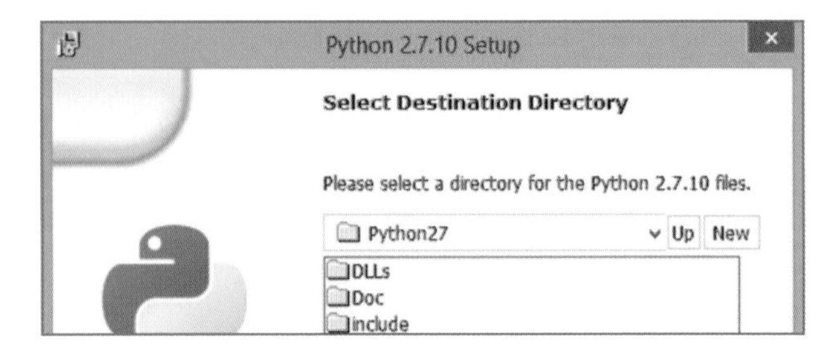

Running an iPad/iPhone Script on a PC

Icons for the two main Python modes of operation are placed on the All apps menu in Windows 10, on the All Apps screen in Windows 8/8.1 and on the Start/All Programs menu in Windows 7.

 Python (command line)  IDLE (Python GUI)

Browsing in Dropbox Using Windows File Explorer

Click the icon shown above right to open the **IDLE** window then select **File** as shown on the next page followed by **Open** from the drop-down menu. Then from the **Open** window shown below, browse to find **Dropbox** and select the required file, in this case dice.py, shown below.

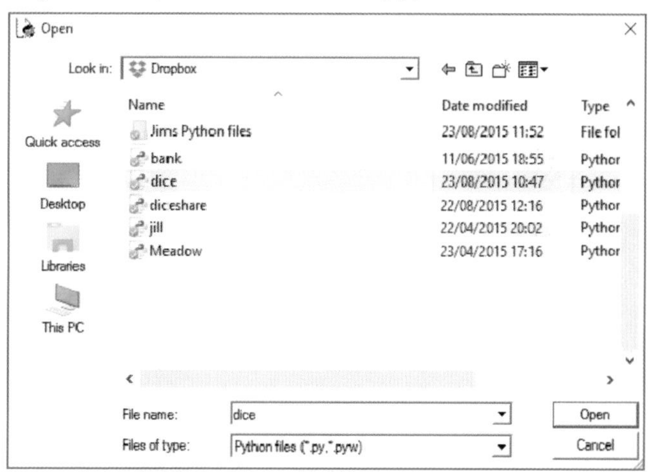

This opens the file dice.py, in the **IDLE Python 2.7 Editor** on the PC, as shown at the top of the next page. From here it can be saved in a folder of your choice on the PC using **File** and **Save As**.... The Windows File Explorer/Manager greatly simplifies the copying of .py files from Dropbox.

The iPad/iPhone Code Listed in the Editor on a PC

After selecting **Open**, as shown on the previous page, the code created in Pythonista on an iPad or iPhone is listed as shown below in the Windows **IDLE Editor** on a PC.

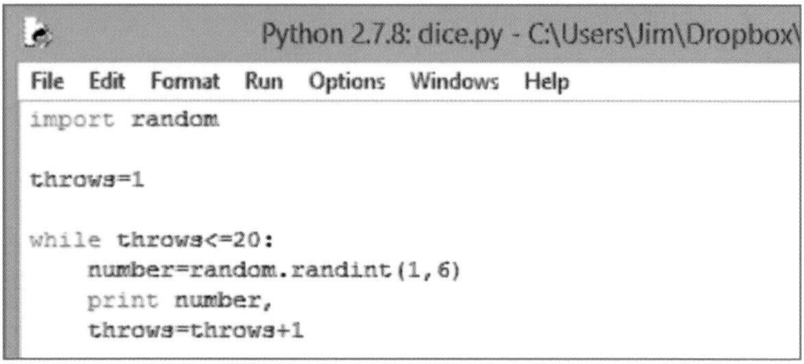

```
Python 2.7.8: dice.py - C:\Users\Jim\Dropbox\
File  Edit  Format  Run  Options  Windows  Help
import random

throws=1

while throws<=20:
    number=random.randint(1,6)
    print number,
    throws=throws+1
```

The Windows Python IDLE Editor

Running the Program in the Python Shell

To run this small program in the **PC Shell** as shown below, select **Run** shown above and then **Run Module**.

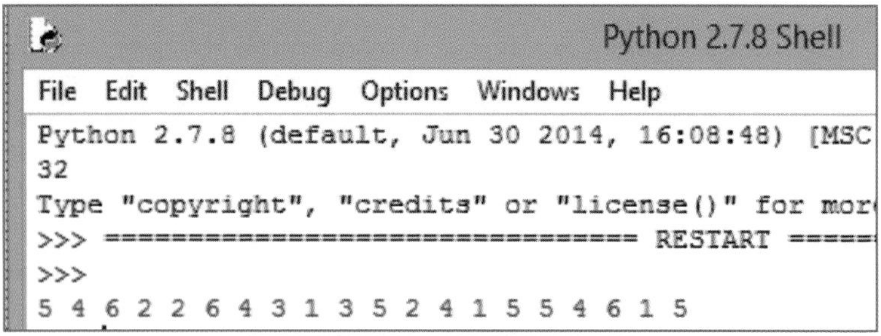

```
Python 2.7.8 Shell
File  Edit  Shell  Debug  Options  Windows  Help
Python 2.7.8 (default, Jun 30 2014, 16:08:48) [MSC
32
Type "copyright", "credits" or "license()" for mor
>>> ================================ RESTART =====
>>>
5 4 6 2 2 6 4 3 1 3 5 2 4 1 5 5 4 6 1 5
```

The Windows Python Shell

As shown above, the program has generated 20 random numbers between 1 and 6, as discussed on page 101 and 102.

Copying a .py File to an Android Device

Install a Cloud Storage System

This section assumes you have copied the required .py file from an iPad or iPhone to Dropbox or a similar cloud storage system. The Android device may need to have Dropbox or Google Drive, etc., installed from the Play Store and be signed into an account.

Install QPython

You also need to install the QPython app from the Play Store. For compatibility with Pythonista, you need to use QPython rather than QPython3. Both QPython and Pythonista are based on the Python 2.7 interpreter.

Method 1:Connect the Android to a PC

Use the battery charger cable to connect the Android to a USB port on a PC. The PC detects the Android like a removable drive and displays the Android Python folder **com.hipipal.qpyplus** in the File Explorer, as shown below.

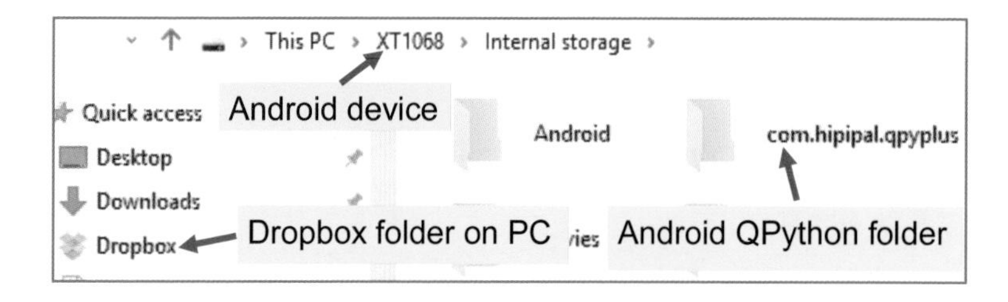

The file dice.py, created on the iPad/iPhone, can now be copied from **Dropbox**, shown above, to the QPython folder on the Android (**com.hipipal.qpyplus** shown above) using drag and drop or **Copy** and **Paste**.

Method 2: Using an Android File Manager

This example again uses the file dice.py, created on an iPad and uploaded to Dropbox. We need to copy it from Dropbox to the QPython folder, **com.hipipal.qpyplus**, on the Internal Storage of the Android.

This can be done after installing the ***ES File Explorer*** app from the Play Store. Then, with the file displayed in Dropbox as discussed earlier, tap the small, very faint arrow in a circle, to the right of the file name, as illustrated (much darker) on the right.

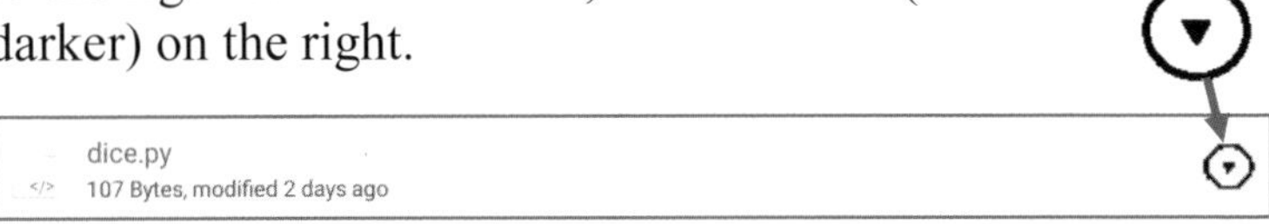

This opens the menu bar shown in part on the right, from which you select **Export**. From the next menu which appears, select **ES Save to...** shown on the right. Then select the QPython folder **com.hipipal.qpyplus** shown on the left below. If you wish, select a ***sub-folder***, such as **projects**, shown on the right below.

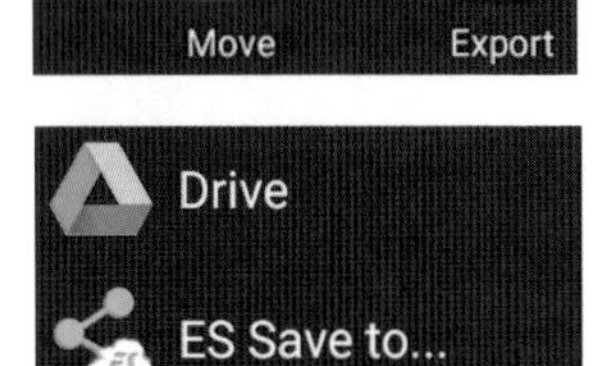

Tap **Select** to save the file in the required folder.

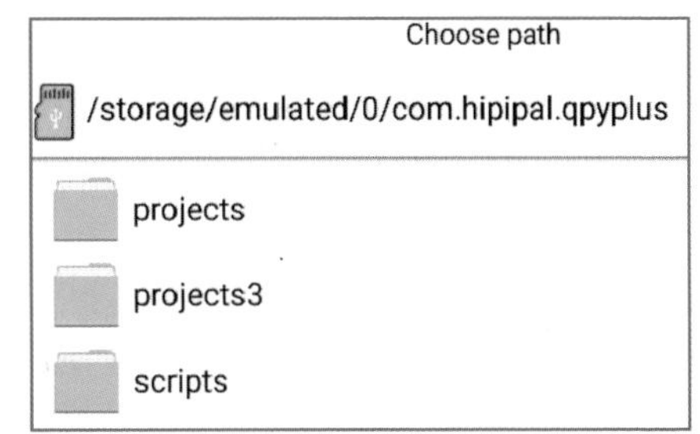